ᵏR

WITHDRAWN

MODERN PHYSICS
IN AMERICA
A MICHELSON-MORLEY
CENTENNIAL SYMPOSIUM

AIP CONFERENCE PROCEEDINGS 169

RITA G. LERNER
SERIES EDITOR

MODERN PHYSICS IN AMERICA
A MICHELSON-MORLEY CENTENNIAL SYMPOSIUM

CLEVELAND, OH 1987

EDITORS:

**WILLIAM FICKINGER
& KENNETH L. KOWALSKI**
CASE WESTERN RESERVE
UNIVERSITY

AMERICAN INSTITUTE OF PHYSICS   NEW YORK 1988

L.C. Catalog Card No. 88-71348
ISBN 0-88318-369-2
DOE CONF-8710320

Printed in the United States of America.

CONTENTS

Prologue vii
    William Fickinger and Kenneth L. Kowalski
National Advisory Committee xi
Acknowledgments xiii
Celebration xvii
    John Schoff Millis

Introductory Remarks 1
Physics at Higher Energies and Smaller Distances: Are There Limits? 6
    Wolfgang K. H. Panofsky
Reminiscences of My Father 22
    Dorothy Michelson Livingston
Atoms, Molecules and Light 26
    Arthur L. Schawlow
The Life of the Stars 49
    Hans A. Bethe
Neutrinos from the Atmosphere and Beyond 65
    Frederick Reines
The Search for Gravitational Waves: Probing the Dynamics of Space-Time 79
    Peter F. Michelson
Experimental Gaze at Nonlinear Phenomena 95
    Albert Libchaber
A Physicist's View of Biology 110
    Ivar Giaever
Strange Insulators, Strange Semiconductors, Strange Metals: High T_c as a
Case History in Condensed-Matter Physics 141
    Philip W. Anderson
Grand Challenges to Computational Science 158
    Kenneth G. Wilson
The Supercollider: Assault on the Summit 170
    Leon Lederman
Is the Whole Universe Composed of Superstrings? 185
    Murray Gell-Mann
SN1987A: The Supernova of a Lifetime 204
    Robert Kirshner
The Discovery and Physics of Superconductivity above 100 K 220
    Paul Ching-Wu Chu
Participants 241

89-2024

PROLOGUE

In Cleveland during the summer of 1887 two faculty members of our progenitor institutions determined, with unprecedented and compelling precision, that the earth's motion had no effect on the behavior of light. This null result was in direct contradiction to the prevailing conception of space-time which held that the motion of the earth through an ether should be accompanied by changes similar to those that occur in the propagation of mechanical waves in material media.

The events of October 30-31, 1987 described in these proceedings were presented by the Physics Department of Case Western Reserve University as part of the Michelson-Morley Centennial Celebration.

The title, *Modern Physics in America,* reflects the principal theme of this symposium which is derived from the clear position of the Michelson-Morley experiment, especially in the United States, at the beginning of what is usually referred to as modern physics. The major goal of the symposium was to celebrate the extraordinary century of American physics that the experiment of 1887 initiated. This is reflected in the program which includes talks by leading physicists in most of the major areas of physics.

In spite of this shift in emphasis away from the 1887 experiment and the people who carried it out, important legacies, particularly of Michelson, were part of the fabric of the two-day program. The circumstance that Albert A. Michelson was the first American to be awarded a Nobel prize in physics is reflected in the strong support of the symposium from the community of American Nobelists in physics. The contributions of these distinguished scientists both as members of the National Advisory Committee for Modern Physics in America and as participants in the symposium are deeply appreciated.

Michelson also pioneered and brought to perfection techniques that were used for astrophysical measurements. It is therefore appropriate that part of the program concerned contemporary versions of such approaches. One of these talks, by Frederick Reines, is based on unusual measurements made with an enormous particle detecting device located in a salt mine just outside Cleveland. These measurements refer to the supernova SN 1987A, described by Robert Kirshner, that evolved from a star named after a Case Western Reserve University astronomer, Nicholas Sanduleak. The talk by Peter F. Michelson on the detection of gravitational waves impinging upon the earth from other parts of the universe represents a different, but no less striking, coincidence.

It was no coincidence, however, that a number of repetitions of the Michelson-Morley experiment were carried out by Edward W. Morley at Western Reserve University and, mainly, by Dayton C. Miller, who was Michelson's successor in the Physics Department of what was then the Case School of Applied Science. The original purpose for doing this was to complete the original 1887 program of repeating the experiment at regular intervals throughout the solar year. This program was finally completed by Miller during 1921-26 in a series of trials in Cleveland and at Mount Wilson. The choice of a mountain location for some of the trials was made in order to test various hypotheses that could account for the sea-level null result but led to altitude-dependent effects.

The initial trials at Mount Wilson indicated the possibility of a small positive effect, although these results were suspected by Miller as specious because of the difficult experimental circumstances. Later trials further reinforced his impression that the sources of errors affecting the experiment were not fully under control. Miller received strong encouragement to continue the Mount Wilson measurements from Albert Einstein and H.A. Lorentz who visited Case in 1921 and 1922, respectively. At stake was nothing less than the experimental underpinning of the relativity principle.

Miller completed his Mount Wilson experiments and published the results which stood at variance with the results of the original Michelson-Morley experiment for more than thirty years. The matter was finally resolved in a paper (*Reviews of Modern Physics* 27, 167 (1955)) by R. S. Shankland, S.W. McCuskey, F.C. Leone, and G. Kuerti who then were all faculty at Case Institute of Technology. Shankland, who followed Miller as Chairman of the Physics Department at Case, led the reanalysis of Miller's Mount Wilson experiment. This group succeeded in ascribing sources of error for all of the previously observed discrepancies and essentially settled the matter.

Robert S. Shankland maintained a keen interest in the Michelson-Morley experiment and its impact on physics throughout his career at the University until his death in 1982. His activity helped to maintain a sense of continuity about these matters from 1887 to the present.

This sense of continuity is enhanced further by the talk "Celebration" given by John Schoff Millis, former Chancellor of Case Western Reserve University, at the opening ceremonies of the Centennial Celebration in April of 1987. This talk contained such unusually engaging personal impressions of both Michelson and Morley that we have included it in the proceedings even though it was not part of the symposium. It also represents the last public address of Chancellor Millis who died a few months after this symposium.

The past was dwelt upon relatively little at the symposium, and then only as a prelude to present developments. The speakers succeeded in bringing the over 1000 participants up to date on some of the most exciting developments in physics. More than a half of these participants were Physics graduate students which led to an air of excitement and anticipation more typical of the artistic performances that normally take place in Severance Hall, the site of the symposium, than of the conferences we usually attend. The speakers did not let them down. The audience response was always spirited, often dramatic.

A unique aspect of the symposium was the number of graduate student participants. It is possible that there have never been that many physics graduate students in one place before. Their camaraderie did much to set the tone and stimulate the speakers. These students, most of whom were bussed in from universities within a radius of a few hundred miles from Cleveland, seemed to be keenly aware of the remarkable educational experience of which they were such a crucial component. The CWRU Physics Department graduate students hosted a Halloween party the first evening for the visiting students after the symposium buffet at the Cleveland Art Museum which further enhanced the prevailing sense of comradeship. The buffet was set in the elegant outer lobby of the museum. The lobby adjoined the exhibit "*Creativity in Art and Science*" that was staged as part of the program of the Cleveland centennial festival "*Light, Space, and Time*." After the buffet, some people returned to Severance Hall to hear a work composed by Philip Glass, "*The Light*," which was commissioned especially for the Michelson-Morley Centennial Celebration, performed by the Cleveland Orchestra.

The remainder of the symposium participants were largely professional physicists including a contingent of CWRU alumni, some members of the National Advisory Committee, and physicists from universities, industrial and government laboratories mainly from within the same geographical circumference that most of the graduate students originated from. Despite the seemingly chauvinistic theme of the symposium, there was a significant number of participants from abroad, largely from Canada and mostly graduate students. The event would not have been quite the same without the cohort of Canadian students. Perhaps next time around it should be called *Modern Physics in North America*, or more optimistically, ... *in the Americas*.

The most difficult dimensions to describe relate to the enthusiasm, optimism, and general good spirit of all of the participants, audience and speakers alike. It is rare to see an intrinsically competitive breed such as the like of physicists in a state of relaxed alertness. The marvelous weather we had for the two days helped a lot as did the beauty of Severance Hall, along with the good food and accommodations that were made possible by the generous funding of our supporters. Also, the many details all seemed to go fairly smoothly due to the extraordinary efforts of all of the people who worked with us. But there was something more to it that is very difficult to verbalize. At the very least, we were all made aware of our extremely good fortune to be part of the richly rewarding human endeavor that the profession of physics represents.

William Fickinger and Kenneth L. Kowalski
□□

Leon M. Lederman   Fermi National Accelerator Laboratory
T. D. Lee   Columbia University
Philip Morrison   Massachusetts Institute of Technology
Yoichiro Nambu   University of Chicago
A. Pais   Rockefeller University
George E. Pake   Xerox Corporation
Martin L. Perl   Stanford Linear Accelerator Center
Isidor I. Rabi *   Columbia University
Norman F. Ramsey   Harvard University
Frederick Reines   University of California at Irvine
Robert C. Richardson   Cornell University
Louis Rosen   Los Alamos National Laboratory
Carlo Rubbia   Harvard University
Robert G. Sachs   University of Chicago
Nicolas P. Samios   Brookhaven National Laboratory
Arthur L. Schawlow   Stanford University
John H. Schwarz   California Institute of Technology
Emilio C. Segrè   University of California at Berkeley
Frederick Seitz   Rockefeller University
Samuel C. C. Ting   Massachusetts Institute of Technology
George H. Vineyard *   Brookhaven National Laboratory
Steven Weinberg   University of Texas
Victor F. Weisskopf   Massachusetts Institute of Technology
Eugene P. Wigner   Princeton University
Frank Wilczek   University of California at Santa Barbara
Clifford M. Will   Washington University
Lincoln Wolfenstein   Carnegie-Mellon University

*Deceased

ORGANIZING COMMITTEE

Co-chairmen:
William Fickinger
Kenneth L. Kowalski

Leslie L. Foldy
William L. Gordon
Thomas H. Moss
Philip L. Taylor

□□

Acknowledgements

Many people contributed to the Modern Physics in America Symposium project, from its inception in the fall of 1985 to the completion of these published proceedings in the spring of 1988. It is impossible for us to acknowledge them all adequately, but we shall attempt to identify those who made particularly noteworthy contributions. We extend our gratitude to them and to everyone else who helped bring this component of the Michelson-Morley Centennial Celebration (M^2C^2) to fruition.

Philip L. Taylor's position as a member of our Organizing Committee obscures his roles as both the originator of the concept of a multi-event festival to celebrate the centennial of the 1887 experiment and as Executive Director of the organization, M^2C^2, which coordinated the staging of these events. His advice and considerable efforts on our behalf were of inestimable value.

Our Organizing Committee's first major task was to define the theme, scope, and potential audience for the symposium. After this was settled they worked very hard with us in developing the program and in its execution. Robert W. Brown, who was an original member of the Committee, provided much in the way of stimulating advice throughout the duration. William L. Gordon, chairman of the CWRU physics department, gave this project his full support and was especially helpful in coordinating the efforts of the departmental staff during the symposium.

The initial and sustained positive response of the Physics community, as represented by the people who joined our National Advisory Committee, was crucial for encouraging us in our endeavor, in attracting funding and speakers, and in providing us with numerous valuable suggestions and wise counsel. Our sole regret is that some fields of indisputably "modern physics" were not adequately represented in the program, principally because of limitations of time. We appreciated the good grace with which these constraints were accepted.

Dorothy Humel Hovorka and Polly Bruner, Chairman and First Vice Chairman, respectively, of M^2C^2, played crucial roles in securing major funding as well as access to the resources and facilities of Cleveland's most prestigious institutions. Their unique contributions throughout the planning stages assured the success of the symposium and especially the social events that accompanied it.

M. Roger Clapp, Vice Chairman of M^2C^2, was vital to the realization of this project through his strong support of our goal of having an unprecedented number of Physics graduate students as a major component of the audience. His foresight and confidence in this idea as an indispensable ingredient for the success of the symposium were essential in making that idea a reality.

Many potential disasters in our planning, staging, and program materials were averted with timely help and suggestions from Joan A. Risberg, the Program Director of M^2C^2. She was ably assisted in much of this by Julie G. Bailey.

The art work for the two Modern Physics in America (MPA) posters and the production of the program booklet were done by Thomas M. Rask of the Publications Office at CWRU. Fred Karaba and his team at the university printing department handled the design, printing and mailing of many thousands of pieces.

William M. Osborn, Director of Media Relations for M^2C^2, was responsible for making the general public, across the nation as well as in Cleveland, aware of the occurrence of the Michelson Morley Centennial and its general significance.

Peggy W. Englehorn was executive secretary for the MPA office. She managed the massive pre-registration and mailing efforts, coordinated all the hotel and travel reservations, and assisted in the planning of the spouse's program.

All of the CWRU Physics Department secretaries, Angeline T. Amato, Jacquelyn A. Hall, Lucille J. Rosenberg, Dorothy M. Straughter, and Margery E. Young, contributed from the beginning of the project through the hectic days just prior to and during the symposium. They were aided in some of these efforts by the wives of several of our colleagues, including Betty Eck, Roberta Farrell, Jean Gordon, and Sarah Taylor.

The person who worked miracles in getting out last minute mailings of several thousand letters was Eleanor Shankland, the widow of our late colleague Robert Shankland. Mrs. Shankland on several occasions arrived to save the day at the head of an army of much appreciated envelope stuffers.

John Procyk, the CWRU Physics Department administrative manager, was responsible for coordinating the work of the secretaries, monitoring our financial matters and our Severence Hall telecommunications. A number of related activities were carried out by George Cadwallader and Sandor Hanzmann.

Virtually all the CWRU Physics graduate students helped with the transportation of the speakers and during registration. Under the leadership of Robert Kusner, they planned and staged the party for the visiting graduate students.

Participation in the symposium by several hundred graduate students was made possible by the provision of charter busses from the principal university centers in the area. It was essential to have contacts at each departure point to help us get the message to the students and to organize the bus operations. Among these special friends were Jonathan Lukin of Carnegie-Mellon, Zoran Pasameda of SUNY at Buffalo, Yizhong Fan at Cincinnatti, Greg Latta at Kent State, Prof. Jean Krisch at U. of Michigan, Paul Rutt at Michigan State, Julie Franklin at Ohio State, Ralph Sherriff at U. of Pittsburgh, Richard Smith at U. of Toronto and Winston Win of Wayne State.

Acknowledgements XV

It is not redundant to call attention to the contributions of the speakers. The polished, edited versions of their talks are pale reflections of the power and presence that were apparent in the actual presentations. The styles of the talks were as varied as the speakers themselves, but they all seemed to reflect a secret conspiracy to generate a steady state of anticipation and every variety of dramatic nuance. There was even a standing ovation after one talk, unusual for an audience of physicists.

The session Chairs, esteemed physicists themselves, provided a number of amusing and insightful comments of their own in addition to giving introductions that provided smooth transitions to the speaker's presentations.

The success of the symposium depended greatly upon its setting in Severence Hall and the efforts of Nancy Groezinger and her staff in helping us exploit this facility in the middle of a three-day concert series by the Cleveland Orchestra.

All of the talks were video- and voice-taped by CWRU staff. Edward Kling directed the three-camera video recording. Charles Hamilton and Michael Kubit, of our Educational Media Department were responsible for the flawless sound system and slide projection. Charles Knox acted as liaison between the speakers and the projectionist: all the slides were projected right-side up!

Todd Harris, a freshman physics major, Jonathan M. Obien, a grad student at Wayne State University, and Suvro Datta, a grad student at Carnegie-Mellon University, took the candid photographs included in these proceedings. Ardon Wilfong supervised the copying and reduction of all the halftone photographs.

Linda Jones, a professional court stenotypist, sat in the front row through the entire symposium to provide us with floppy disks of each talk. (As a matter of note, the text on these floppies was checked against the audio recordings, edited, and sent to the speakers. Most of the papers included in these proceedings are thus speaker-edited versions of the original spoken versions.)

Evalyn I. Gates, a CWRU physics graduate student, participated in several stages of the editing of these proceedings. She helped organize our editing routines: she worked on the first-stage floppies and on the final formating of the camera ready copy.

The computer graphics work was done by three CWRU undergraduates and Macintosh enthusiasts: Charles Fernandez, Robert Klepfer, and Paras Shah.

Robert B. Akins, a CWRU physics graduate student, and one of our colleagues, Donald E. Schuele, generously gave us access to their Macintosh system. This entire volume was produced using the new desktop publishing software: FULLWRITE PROFESSIONAL of Ann Arbor Softworks, Inc.

Our A.I.P editor, Rita Lerner, provided cogent advice and patient understanding of our deadline violations.

This symposium was made possible entirely through the generous support of The 1525 Foundation and the General Electric Foundation.

The participation in the symposium by physics faculty from four-year colleges was subsidized by a grant from the National Aeronautics and Space Administration.

W.F. and K.L.K.

□□

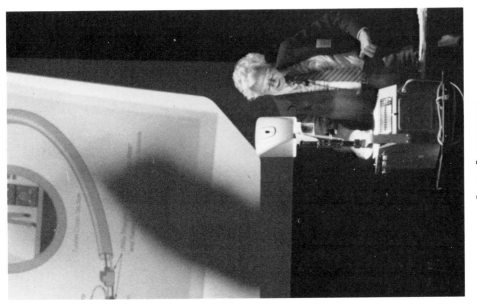

Leon Lederman

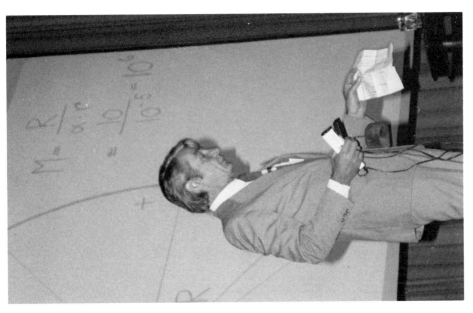

Ivar Giaever

Murray Gell-Mann

Dorothy Michelson Livingston

BROOKHAVEN'S GOLDHABER, SLAC'S PANOFSKY AND FERMILAB'S LEDERMAN

ALBERT J. LIBCHABER   WILLIAM FICKINGER

KENNETH L. KOWALSKI

MICHAEL E. FISHER

AFTERNOON BREAK
IN THE GRAND FOYER
OF SEVERANCE HALL.

MICHELSON–MORLEY
BANNERS FLY OUTSIDE
SEVERANCE HALL.

SCALE REPLICA OF MICHELSON–MORLEY INTERFEROMETER IN SEVERANCE FOYER.

LETTER FROM MORLEY TO HIS FATHER DESCRIBING THE PROPOSED EXPERIMENT.

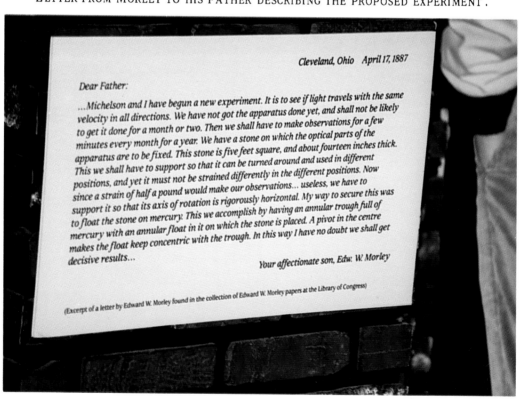

Cleveland, Ohio   April 17, 1887

Dear Father:

...Michelson and I have begun a new experiment. It is to see if light travels with the same velocity in all directions. We have not got the apparatus done yet, and shall not be likely to get it done for a month or two. Then we shall have to make observations for a few minutes every month for a year. We have a stone on which the optical parts of the apparatus are to be fixed. This stone is five feet square, and about fourteen inches thick. This we shall have to support so that it can be turned around and used in different positions, and yet it must not be strained differently in the different positions. Now since a strain of half a pound would make our observations... useless, we have to support it so that its axis of rotation is rigorously horizontal. My way to secure this was to float the stone on mercury. This we accomplish by having an annular trough full of mercury with an annular float in it on which the stone is placed. A pivot in the centre makes the float keep concentric with the trough. In this way I have no doubt we shall get decisive results...

Your affectionate son, Edw. W. Morley

(Excerpt of a letter by Edward W. Morley found in the collection of Edward W. Morley papers at the Library of Congress)

SHAFT OF MICHELSON–MORLEY FOUNTAIN AND CRAWFORD HALL ON CWRU CAMPUS.

AGNAR PYTTE, DOROTHY HUMEL HOVORKA, PHILIP L. TAYLOR, WILLIAM FICKINGER.

LESLIE L. FOLDY

MILDRED S. DRESSELHAUS

ROBERT HOFSTADTER

MAURICE GOLDHABER

ARTHUR L. SCHAWLOW

FREDERICK REINES

HANS A. BETHE

PETER F. MICHELSON

ROBERT KIRSHNER

KENNETH G. WILSON

CHING WU PAUL CHU (RIGHT)

PHILIP W. ANDERSON

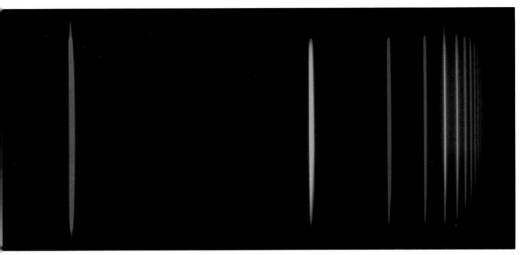

FIG. 2.

ARTHUR L. SCHAWLOW

FIG. 3.

FIG. 18.

FIG. 4.

FIG. 25.

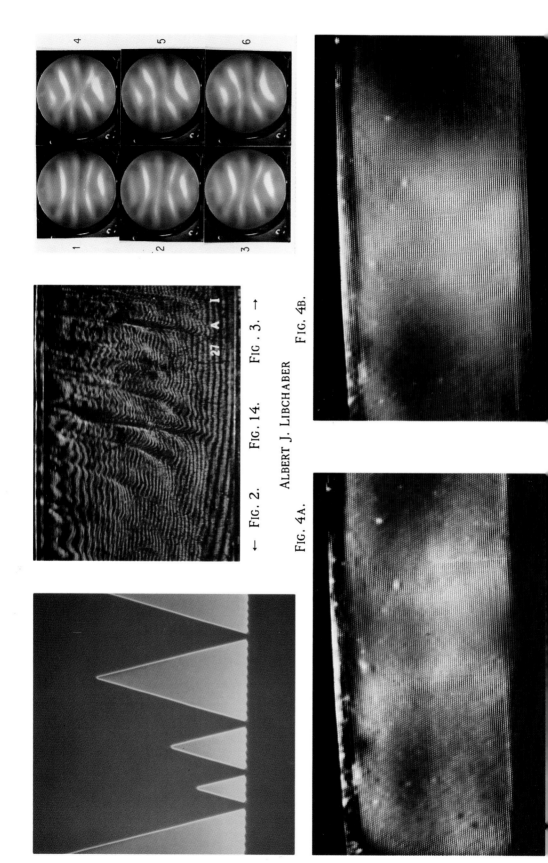

FIG. 4A.

FIG. 4B.

FIG. 2. ←

← FIG. 3. →

FIG. 14.

ALBERT J. LIBCHABER

Celebration

John Schoff Millis

CASE WESTERN RESERVE UNIVERSITY

Cleveland, Ohio 44106

This evening we have begun the celebration of the Centennial of the Michelson-Morley Experiment, which was conducted in 1887 only a few hundred yards from where we are now gathered, by two distinguished and creative citizens of Cleveland: Albert Abraham Michelson, Professor of Physics at the then Case School of Applied Science and Edward Williams Morley, Professor of Chemistry at Western Reserve University. Some scholars listed the Michelson-Morley Experiment as one of the most crucial experiments in Man's long search for understanding of the world in which we live and in the record of the thinking, creating and accomplishing beings that we are.

I leave the discussion of the relative importance of the Michelson-Morley Experiment to those far better qualified than I. The Symposia scheduled in the ensuing months will spend much time in evaluating the work of Albert A. Michelson and Edward W. Morley. I am confident that they will find that the contributions of those two Clevelanders were of significance to the century which all must recognize as the Golden Age of Science.

I wish to speak about the lives and personalities of the two men whose joint scientific accomplishment we are celebrating.

Albert Abraham Michelson was born in Strzelno, Poland in 1852 and came as a child with his family to the United States to settle in the mining town of Murphy's Camp, California. Most of his earlier education was received in San Francisco where he lived with relatives. Having demonstrated unusual capabilities he gained admission, albeit with some difficulties with formal requirements, to the U.S. Naval Academy at Annapolis. Following graduation and a tour of duty at sea he returned to the Academy as instructor in the physical sciences. His life-long interest in the velocity of light began in 1878 with his first measurements in that year.

In the period 1880 to 1882 he pursued the graduate study of physics in several German Universities particularly with the famous scientist Helmholtz in Berlin. He came to Cleveland in 1882 to join the faculty of the then Case School of Applied Science. In 1889 he moved to Clark University and to the University of Chicago in 1895 to head the department of Physics as that University began its illustrious career.

Throughout his career, Mr. Michelson never wavered from his absorbing interest with light and its properties. He was famous as an experimentalist who demanded the utmost accuracy of his apparatus and experimental procedures. He was highly skilled as a designer and manufacturer of optical devices. His life was given to developing apparatus of greater precision and to ever reducing the errors in his observations. One must think of him as a perfectionist.

It was my great good fortune in the 1920's to be a student of Physics at the University of Chicago and to be a member of two courses taught by Albert Abraham Michelson in the closing years of his distinguished career as Professor of Physics. I remember clearly the first meeting of a course entitled "Physical Optics." The class was assembled in the second floor lecture room of the Ryerson Physics Laboratory. I awaited with excitement the entrance of the world famous scientist, the first American to win the Nobel Prize in the sciences. When Professor Michelson entered the room, I was struck by the carefulness of his dress and the neatness of his grooming. The idea flashed through my mind that here was a man who believed that teaching was a responsibility of highest importance and required the most careful preparation.

Professor Michelson proceeded to present a superb lecture. His organization was precise and clearly evident. His languages was well chosen and revealed a scholar's command and respect for the mother tongue. In short a class taught by Professor Michelson was a rich and rewarding intellectual experience. In all the classes I attended, in all the public lectures I heard, Professor Michelson certainly was the finest teacher I have ever known.

Over the years I learned more about Albert Abraham Michelson, the man. The route from my house on Kenwood Avenue to Ryerson Laboratory led me past his home on 58th Street several times a day. On several times, perhaps many times, I heard Professor Michelson practicing the violin. Though I am no music critic, his playing seemed that of an accomplished practicing the violin. Though I am no music critic, his playing seemed that of an accomplished musician. More competent judges have acclaimed him as highly gifted.

Other memories of Professor Michelson complete my picture of a man of many talents, ever striving for perfection. I recall viewing an exhibition of his paintings, the fruit of another of his many talents. I have watched him at the Quadrangle Club demonstrate his skill at billiards and at bridge and have seen him playing tennis on the campus courts.

Albert Abraham Michelson was a man of many gifts and many talents. He excelled as an experimentalist, as a teacher, as a musician, as a painter. In him the fire of creativity in science, in music, in art, are manifestations of that greatest gift to mankind, the intelligence, the spirit, the free will to be creative.

I never had the privilege of meeting Edward Williams Morley, for he retired from his position as Professor of Chemistry at Western Reserve University in 1906 and returned to West Hartford, Connecticut, where he died in 1923. Fortunately, due to the efforts of Professors Frank Hovorka and Harold Booth, the University Archives has a large collection of Professor Morley's papers and correspondence and many letters from his friends and scientific colleagues. From them I have gathered a most interesting picture of the life, career, and personality of a remarkable man and a master scientist.

At first reading, the life career, and personality of Edward Williams Morley appear to have been very different from that of Albert Abraham Michelson. Morley was born to strict Congregationalist parents in Newark, New Jersey in 1838. His father was a minister and the son seem destined for the Ministry. Because of frail health, he studied at home until he was nineteen when he entered Williams College in 1857. Following graduation he entered Andover Theological Seminary in 1861. He began teaching both Theology and Science in South Berkshire Academy in 1866.

In 1867 he became the Minister of the Congregational Church of Twinsburg, Ohio. In that same year, the Trustees and Faculty of Western Reserve College decided to add a Department of Science. It is reported that a deputation of Trustees and Faculty walked from Hudson to Twinsburg to interview the young clergyman whose interests and talents in science had become known in the region. The result of the interview was that Mr. Morley was offered and accepted the appointment as the first Professor of Science in Western Reserve College.

In the early years, Professor Morley taught Physics, Chemistry, Mathematics, Mineralogy and Botany. He even found time to give lectures at the Medical School in Cleveland and to play the organ at Chapel exercises. In 1882 Western Reserve College moved to Cleveland and became Western Reserve University and Morley was able to concentrate his energies upon Chemistry, which had become his chief but not exclusive interest and concern. He first gained national and international acclaim for his precise measurement of the properties of gases. These researches on the density of oxygen and hydrogen and the ratio of their atomic weights, published in 1895 remain unequaled for accuracy and thoroughness.

The best way I can give you some feeling and knowledge about Mr. Morley, the man and the scientist, is to quote from letters written by his friends and colleagues. In a letter in 1921, A. W. Smith, the greatly admired and much loved Professor of Chemistry at Case School of Applied Science wrote the following sentences.

"During the early years of my life at Case School I saw a great deal of Professor Morley and was with him perhaps as much as any of his friends. Not only had he a mind of rarest quality, but a kindliness equally great. This was evidenced frequently in my case with the assistance he was always ready to give a young man just starting in on his life work"..."His interests were not confined to Chemistry alone; but while this was his specialty, his interests were as broad as science itself."..."His interest in music also shows the breadth of his mind, as he enjoyed the best music keenly and was a good performer on the pipe organ."

I further quote several sentences from a letter written in the same year by Dayton Miller, the much honored Professor of Physics at Case School in the early decades of this century.

"Of the 52 papers published by Professor Morley, 23 are in pure physics, 27 in pure chemistry, 2 in general subjects. He had 4 collaborators in scientific work, all physicists; three papers with Michelson, one with Rogers, one with Brush, and seven with Miller."

"Professor Morley was one of the three or four greatest men I have known in my whole scientific acquaintance. He was great because he was so widely and deeply informed because he philosophized so truly, so scientifically, without prejudice, and his conclusions were precise and concise."

I hope that I have given you some feeling and understanding of these two remarkable men. Their early lives, education, and careers were very different. Michelson was already committed to his life-long absorption with light as an undergraduate at Annapolis. He pursued that passion throughout his life. Morley began his career as a clergyman and became a scientist almost by accident. He began with an interest in all science and moved slowly to a concentration in both Chemistry and Physics. Light was only one of many physical phenomena which caught his interest, challenged his intellect, and demanded his rare skill as an experimentalist.

On the other hand both men were famous for the precision of their scientific thought and their insistence upon the perfection of their apparatus and experimental procedures. Both were men of many talents, both were talented musicians. Both were known as superb teachers.

This evening begins the celebration of the Centennial of a great scientific achievement which occurred here in what we now call the University Circle, the remarkable educational and cultural asset of Cleveland. But our celebration is much more than that of a great scientific achievement. It is the celebration of the lives and contributions of two great men. In a larger sense, it is the celebration of mankind, his intellect, his unquenchable spirit which leads us toward truth, and his free will which drives him to so live and work that civilization may advance and we may be more wise and more humane.

An address by John Schoff Millis, Chancellor Emeritus of Western Reserve University, on April 24, 1987, Severance Hall, Cleveland, Ohio. Dr. Millis passed away in January 1988 in his eighty-fourth year.

□□

Introductory Remarks

WILLIAM FICKINGER:

Good morning everyone. Welcome to Cleveland, welcome to Case Western Reserve and welcome to the Michelson-Morley Centennial Symposium. My name is William Fickinger, and I'm a high energy experimental physicist. This year I'm co-chairman of this Symposium. My partner on this project is Ken Kowalski, who will be speaking to you later today. I would like briefly to thank those who have made these events possible. First I'd like to thank our distinguished National Advisory Committee, seventy outstanding physicists who have generously associated themselves with this centennial. Among these are twenty-two Nobel Laureates.

Next, I would like to thank the General Electric Foundation and the 1525 Foundation who agreed with us that this was a fine way to mark the Centennial, and who had confidence in our plan, and I must say a great sense of adventure. The 1525 Foundation is a Cleveland private trust with traditional ties to the University. In addition, NASA has provided funds to subsidize participation by physicists from four-year colleges.

We have about eight hundred pre-registered participants, including about 450 physics graduate students. This morning, an additional 200 or so physicists and other friends have registered. Then I'd like to thank our speakers and point out that they have agreed to join us today because we have promised them you, this extraordinary audience of students and teachers of physics. Finally I wish to thank you, the audience, for joining with us in this extraordinary celebration.

We'll spend very little time in ceremonial talks and housekeeping announcements. However, I must explain one thing: our schedule is very tight, especially at the end of each afternoon. The Cleveland Orchestra will perform in this wonderful hall both tonight and tomorrow night. Therefore, we must be out of this hall at five P.M. each day. Please remember that if we seem a little brusque when we adhere to our schedule. It's my pleasure to introduce to you the Executive Director of the Michelson-Morley Centennial celebration. Professor Philip Taylor is a theoretical condensed matter physicist. He will tell you a bit more about the celebration this year. Phil.

PHILIP L. TAYLOR:

Thank you, very much. As physicists we are only too well aware of the difficulty of communicating the excitement of our subject to a wide audience, and it is a wonderful opportunity to have a thousand enthusiastic people gathered together in this way.

Normally when you say you are a physicist the response you get from the general public is not always flattering. Sometimes they think you're a gymnast; sometimes you meet someone at a cocktail party who will ask what

you do, and you say, "I'm a physicist". There are generally three responses
you get. One is, "Oh, I was never any good at that". Then you have to think
of something that doesn't sound patronizing in response. Sometimes you'll
meet someone who'll say, "Physics? It was a physics professor who gave me
the C+ that kept me out of medical school", and he cracks his knuckles omi-
nously, "and that's why I went into professional wrestling!"

But, perhaps the most common experience is that when you say you're a
physicist, they reply, "Oh, a physicist", and they look at their fingernails
for a while, and then walk away and start talking to somebody else.

That is really what makes today's event all the more remarkable. We are
now part of now what is probably the biggest celebration of a single scientific experiment that has ever occurred in the history of the world. Our
commemoration of the Michelson-Morley experiment is not confined to the
meeting we're having now: we are part of a Centennial Celebration that has
sponsored over fifty events and that has run from April until its the culmination this week. We have put light sculptures on our tallest buildings. We
have commissioned musical compositions, including the Philip Glass work,
"The Light", that was premiered last night on this very stage. We've produced books, museum exhibits and lectures. In fact, the whole town has become enlivened and excited by a physics experiment, and not even a new
one, but an experiment that is a hundred year old.

Now this great festival that we are having has taken a huge amount of
organization. The major credit for this goes to a dynamo of a lady who is
the Chairman of the Michelson-Morley Centennial Celebration. I call on
Dorothy Humel Hovorka.

DOROTHY HUMEL HOVORKA:

Thank you Professor Taylor. He is very modest, but also extremely creative, for it was he who conceived the idea for a celebration of Science, the
Arts and the Humanities. Of course he never realized that we would carry
his idea as far as we did. We are most grateful to him.

A very warm welcome to this Modern Physics in America Symposium. As
you heard Professor Taylor say, it is the culmination of "M^{2}C^2", our six-
month Michelson-Morley Centennial Celebration. To give you special
greetings on behalf of your host institution is Agnar Pytte, who earlier this
month was inaugurated as President of Case Western Reserve University.

A Phi Beta Kappa graduate of Princeton University, Dr. Pytte received the
AM and PhD degrees from Harvard University. President Pytte came to us
from his position as Provost of Dartmouth College, where he had occupied
the following posts: Chairman of the Department of Physics and Astronomy
and Dean of Graduate Studies. Dr. Pytte is a specialist in theoretical physics
and nuclear fusion. He was a member of Princeton's Project Matterhorn.
His other research appointments have included a National Science Foundation Faculty Fellowship at the Universite Libre in Brussels, and an appointment as a Visiting Fellow at Princeton's Plasma Physics Laboratory. He is
co-author of a book, " *The Structure of Matter*", and has published exten-

sively on theoretical physics. It is my very great pleasure to present Agnar Pytte, the president of Case Western Reserve University.

AGNAR PYTTE:

Thank you, Dorothy. It is indeed a double pleasure for me, as President of the University, and as a physicist, to welcome you this morning to this wonderful symposium. Since I had no part in the planning of the symposium, I think I can boast about it without sounding too boastful. I am especially pleased to have so many graduate students come to us from Toronto, Buffalo, Pittsburgh, Columbus, Cincinnati, Ann Arbor, East Lansing and many other places. I think that is a wonderful thing for this University. As has been said already, this weekend really is a culmination of a six-month long celebration of the Michelson-Morley experiment, which has involved not just this University, but a dozen other cultural and educational institutions in Cleveland.

The Michelson-Morley experiment deserves a commemoration of this type; it was a remarkable experiment. I still remember fondly working with a Michelson interferometer in my senior year at Princeton back in 1952. I was in awe of the remarkable precision that one could get with that instrument. That it was developed here in Cleveland one hundred years ago is really something to admire.

What a century it has been from 1887 to 1987. It has been a century of Modern Physics, and that's what we are celebrating here today. That is what the speakers will talk about. 1887, of course, is also known for the Hertz experiment: the first transmission and reception of electromagnetic waves in the laboratory. Einstein, I believe, was eight years old when Michelson and Morley performed their famous experiment. And we know what followed.

I've often thought about the way we present the Michelson-Morley experiment and Einstein's revelations to our students. Usually we present three principles and then we say they're inconsistent. And this, of course, is what puzzled Michelson and Morley. One principle is that the speed of light is the same in all reference frames. That's what Michelson and Morley discovered, albeit reluctantly. And then we have the so-called principle of relativity, which was the one thing Einstein refused to give up; that is that the laws of physics are the same in all inertial frames. The third principle is essentially Newton's Laws, and the three are inconsistent. Now, the one thing that Einstein did not want to give up was the principle of relativity, that the laws of Physics are the same in all inertial frames, and, of course, the constancy of the speed of light was a fact discovered here a hundred years ago.

I have often wondered what would have happened if somebody had asked the simple question: is there a maximum speed in nature? That's the kind of question that has one of two answers, so a priori you have a fifty percent chance of being right. The same could be said of Democritus when he asked the question, is there a smallest unit of matter or is it infinitely divisible: yes or no. Again, he had a fifty percent chance of being right. But if there

is a maximum speed in nature, and if you believe in the principle of relativity, that the laws of physics are the same in all inertial frames, then that
maximum speed has to be the same in all inertial frames.

And of course, that is where we ended up, but that took a long time. In
any case it's been a remarkable century of physics, and we are going to
hear about that today from our distinguished speakers. And it all started
one hundred years ago with the remarkable Michelson-Morley experiment.

I have one other very gratifying thing to announce this morning. Case
Western Reserve University has just received a one million dollar gift from
Mr. and Mrs. Roger Clapp to endow a chair in Physics, which will be
known as the Albert A. Michelson Professorship. On behalf of the University, and I'm sure on behalf of the Faculty and the rest of you, I extend my
deepest gratitude to Roger and Ann Clapp. Of course, Roger and Ann Clapp
are well known to all of us at the University. They have served, both of
them, on the Board of Trustees. Roger Clapp chaired the Board of Trustees
from 1975 to 1979. He was also at one point a member of the Faculty of Case
Institute of Technology, and he is the retired chairman of the Lubrizol Corporation. I am, indeed, very pleased to be able to announce, during this
Michelson-Morley Centennial Celebration, that we will have a permanent
endowment for the Albert A. Michelson Professorship.

There is one other thing that I want to say. This remarkable celebration
that has gone on now for six months, and involved 12 other educational and
cultural institutions in this area, could not have happened without the
leadership that was mentioned earlier. I want especially to thank Dorothy
Humel Hovorka, the Chairman of the Michelson-Morley Centennial Celebration and Polly Bruner, the first Vice-Chairman of the Celebration; Phil
Taylor, the Executive Director, and Joan Risberg, the Program Director. I
believe that concludes the welcoming ceremonies and we can turn to the
first session.

It is with the greatest of pleasure that I introduce the first speaker,
Wolfgang Kurt Herman Panofsky who was born in Berlin, Germany in
1919. He came to the United States in 1934. He got his Bachelor's Degree at
Princeton in 1938 and his PhD at Cal Tech in 1942. Of particular interest to
this audience will be the fact that he received an honorary Doctorate of
Science from Case in 1963. He was at Berkeley from 1945 to '51, and since
then, except when he travels around the world, his home has been Stanford
University, where he is a professor. Since 1962 he has been the Director of
the High Energy Physics Laboratory and the Stanford Linear Accelerator
Center. It takes a special person to be selected to head one of the great national laboratories; not only a comprehensive knowledge of physics, but
also extraordinary judgment. It is one of the most difficult jobs that anyone
can imagine. And Professor Panofsky has carried out his duties with great
distinction.

He is a winner of too many awards to list, but I will mention just a few.
He has received the Ernest O. Lawrence Prize, the National Medal of
Science, the Franklin Medal, and the Enrico Fermi Award. He is a member
of the National Academy of Sciences. But to me, he will always be best

known in another connection. When I was a graduate student in physics at Harvard in the early fifties, there was a book that I absolutely treasured. I don't even remember the exact title of it, because it was never known by it's title, it was simply known as Panofsky and Phillips. And I still treasure that book.

It's a great pleasure for me to be able to introduce to you Professor Wolfgang Panofsky, who has just flown in from Moscow. Professor Panofsky.
□□

Physics at Higher Energies and Smaller Distances:
Are there limits?

Wolfgang K. H. Panofsky

STANFORD LINEAR ACCELERATOR CENTER

Stanford, California 94305

The physics at higher energies and smaller distances are synonymous, since the Heisenberg uncertainty principle requires that large momentum transfers are required to explore the nature of matter in its smallest dimensions. Yet recent progress in high energy physics has linked our increased understanding of physics at the smallest distances also with the physics of the cosmos: there has been a major converging of interest between astrophysics and elementary particle physics.

In talking about limits of physics at higher energies I will principally discuss the limits imposed on the machines which produce charged particle beams at such energies and the limits in using them. Although I will concentrate on the issues associated with the attainment and utilization of such energies, many of the remarks apply also to experimental pursuits in other areas of science. Elementary particle physicists are by no means the only ones who are finding that the tools of their trade are pacing scientific progress.

Elementary particle physicists studying nature at the smallest distances are largely, but not entirely, dependent on charged particle accelerators as a source of high energy collisions. There are productive activities in the field not using such tools. These are principally experiments using cosmic radiation as a source of particles and experiments deep underground to study either the interactions of neutrinos, which penetrate to such detectors, or to search for the existence of decay processes of particles such as the proton which heretofore has been believed to be stable.

Then there are experiments where extremely high precision is substituted for high energy. Such experiments can be powerful if the existence of "virtual" processes at high energy can have demonstrable, but very small effects at low energy. Naturally, this approach is meaningful only if theoretical understanding is sufficiently exact, as is the case in respect to Quantum Electrodynamics (QED).

While such non-accelerator approaches to physics at smallest distances are important, their recent impact on increased knowledge in this field has not been large. The looked-for decay of the proton which has been predicted by theories of the unification of the strong, weak and electromagnetic interactions has not as yet been found. The limit on proton lifetime has been pushed beyond 10^{33} years. This means that detectors containing at least 10^{35} protons and therefore weighing at least 160,000 tons have to be constructed if, even assuming 100% efficiency, 100 such decays are to be

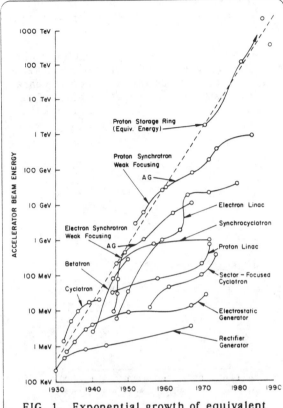

FIG. 1. Exponential growth of equivalent accelerator energy in the laboratory vs. time.

observed per year. Precision electrodynamic experiments have increased our confidence that QED is very precisely correct, but have not uncovered basically new physics. Thus practical limits apply even to such non-accelerator experiments.

Similarly the problem with cosmic rays is that the incident flux of particles decreases drastically with energy. For instance, the cosmic ray flux of particles corresponding to a collision energy to be produced by the Superconducting Super Collider (SSC) (which will be discussed in a later session) corresponds to something like one particle per square kilometer per year - - not a very useful flux for experimentation. I hope, therefore, to be forgiven if in further discussion I will restrict myself to high energy particle accelerators and colliders.

The concept of fundamental limits implies that laws of nature prevent attainment of the requisite parameters. Beyond that one can talk about limits where the data rates resulting would no longer be compatible with the human ability to draw conclusions in a reasonable time, or one can talk about economic limits. The latter, that is economic limits, have been the focus of much discussion when proposals for some of the new projects, appearing extravagant to some, are introduced. However, the direct cost of science in our economic life is still sufficiently small that one has to agree that economic limits are largely in the eyes of the beholder. I would like to remind you of the joking suggestion made in the early 1950's by Enrico Fermi that an accelerator should be built in the vacuum of outer space as a complete ring encircling the earth. I once had a discussion in 1956 in the Soviet Union with Professor Artsimovich, the great plasma physicist. At a lag in the conversation Professor Artsimovich asked me whether I had made a cost estimate of Fermi's proposed machine. I made a quick calculation and answered that "the sum of the U.S. and Soviet military budgets would pay for it in two years." Professor Artsimovich changed the subject.

The growth pattern of particle accelerators measured in terms of energy has been exponential, starting in the 1930's. This is shown in Fig. 1, a chart which we owe to Professor Livingston.

This spectacular growth has been attained not by simply building things larger but by a recurrent series of innovative ideas: whenever one technology appeared to run out of steam a new one was invented, and in consequence the overall exponential growth pattern is really an envelope of saturating growth curves of discrete technologies. A consequence of this series of technological innovative ideas has been the dramatic decrease in cost per unit energy attained. As the energy of high energy particle accelerators has increased by 7 orders of magnitude or so, the cost per unit energy has actually decreased by more than 5 orders of magnitude. This is illustrated in Fig. 2. As a result the cost growth of individual new installations ranging from the electrostatic generators in the 1930's to the proposed Super Collider is 1 or 2 orders of magnitude. Yet all exponentials, be they the energy

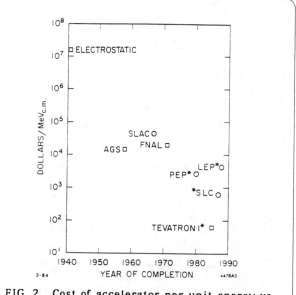

FIG. 2. Cost of accelerator per unit energy vs. time.

of accelerators or colliders, the growth of capabilities of a computer, or the increase of human populations on earth must stop somehow, and it is therefore appropriate to examine the nature of such limitations.

The initial growth of accelerators and colliders involved increasing the energy of particles within beams striking stationary targets. What the physicist is actually interested in is, of course, not the energy of the incident particles but the energy which is available in elementary particle collisions. If one considers the center-of-mass motion of the system of the incident particle and the particle which it is striking, then some of the energy of the bombarding particle is spent in that of the common center-of-mass motion and only a fraction is available for the collision. As energies become very high, relativistic mechanics shows that this fraction of the total energy available for the collision energy in the center-of-mass frame increases only as the square root of the incident energy. Therefore the attention of physicists has shifted from stationary target accelerators to colliding beam machines, or colliders in short, in which beams are produced which collide head-on, or almost head-on, with one another. Here the sum

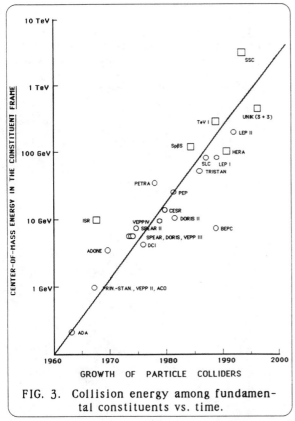

FIG. 3. Collision energy among fundamental constituents vs. time.

of the energies of the colliding particles and the collision energy are identical. Thus the frontier has largely shifted to such colliders and my remarks will principally be addressed to them.

The charged particles used in high energy colliders built for elementary particle physics are either protons or electrons. Protons are now known to be composite particles consisting of quarks bound by gluons, and therefore the total energy of the proton is divided among the energy carried by these constituents. The elementary particle physicist is naturally interested in collisions among those particles which are (at that epoch in time) the most fundamental. Therefore in comparing the relative merits of electron and proton machines the energy of proton machines must be derated by perhaps an order of magnitude relative to that of electron colliders in comparing their respective reach into the unknown. Therefore we can chart progress better by plotting the collision energy between elementary constituents rather than the beam energies; this process converts Fig. 1 into a new chart shown in Fig. 3; note electron and proton energies now track one another by this measure.

But from the point of view of useful physics the comparison between electron and proton colliders is more profound than that. As we go to higher and higher energies one wishes to discover and examine the properties of fundamental objects of heavier and heavier masses. In turn, the cross section governing the production of such massive states is expected to vary inversely with the square of the energy. This expectation is based on the fundamental principle of quantum mechanics known as unitarity, which in essence implies that an interaction process between two particles cannot involve more than an effective area given by the square of the wavelength representing the incident particle. The resultant expected energy variation of cross section is difficult to circumvent. Figure 4 illustrates the general pattern. Here I show you how the $e^+ e^-$ cross section for the production of massive particles as a function of energy. The basic variation goes as one over the square of the energy. The behavior also applies to proton collisions.

For electron colliders this energy variation has the consequence that the cross sections both of new processes of interest as well as those of competing, less interesting processes vary inversely to the square of the collision energy. As a result what we call the luminosity of the collider, that is the data rate per unit of interaction cross section, has to increase as the square of the energy if the data rate is to remain constant. Such a luminosity variation with energy is thus a fundamental design requirement. However, as Fig. 5 indicates, past colliders have failed to obey this rule. We can legitimately ask why this is so and whether future technological developments might correct that situation. Can the accelerator builders provide a luminosity which will keep up with the $1/E^2$ drop in cross sections?

For a proton collider the implication of these energy variations is doubly serious. Not only must the luminosity increase proportional to the square of the collision energy, but the ratio of interesting processes to the less interesting processes is

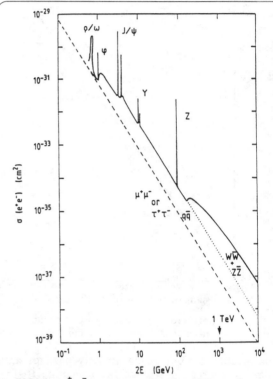

FIG. 4. e^+e^- cross section for the production of massive particles vs. energy.

expected to degenerate, in contrast to what is expected of an electron collider. The reason is that the total cross section for proton-proton interactions is roughly energy independent and in fact increases slowly (logarithmically) with energy. The reason is that the proton, being a composite system, permits interactions to occur both among individual constituents and collectively due to interactions among the constituent quarks and gluons. High energy proton-proton collisions result in a wide range of "hardness" (momentum transfer) of collisions among the fundamental constituents. Thus proton colliders of increasing energy impose an ever-increasing burden on data analysis aimed at extricating the interesting, i.e. "hard" collisions from the background of other processes.

In contrast to this data analysis situation, the accelerator design principles of proton colliders which have been successfully demonstrated at the Tevatron at Fermilab permit extension with confidence into energies now

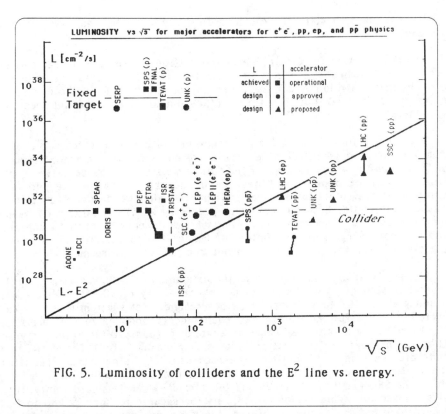

FIG. 5. Luminosity of colliders and the E^2 line vs. energy.

considered for the SSC and probably much beyond, limited primarily by complexity and cost.

The situation in respect to electron-positron colliders is precisely the reverse. Here, as we have just discussed, signal and background vary roughly together as a function of energy; therefore the analysis problem at higher energies of extricating important new discoveries from background should remain at roughly the current level. However the colliders which have recently been enormously successful in this field have been of the storage ring type and this technology is not practically "scalable" beyond the current installations. In such a machine the electrons and positrons are magnetically guided in closed orbits and the particles are brought into collision at a number of interaction points. The problem is that electrons and positrons confined to such orbits radiate electromagnetic radiation (known as synchrotron radiation), producing an energy loss per turn which increases with the 4th power of the energy divided by the radius or the orbit. This relationship forces the designer to increase the machine radius more rapidly than linearly with energy; in fact a quadratic relationship is about optimum. As a result the circumference of the electron-positron storage ring (LEP) at CERN in Geneva, designed to operate up to a collision energy of 200 GeV, is about 27 kilometers. As noted above, this machine, considered by most to be the last attack on the high energy frontier with electron-positron storage rings, is not expected to attain a luminosity adequate to maintain the data rate produced by the lower energy storage rings.

The other option for electron-positron colliders is to use "linear colliders". Here beams are accelerated in two linear accelerators and are brought into collision. Such machines do not exhibit the limits of storage rings just discussed, but they invoke other fundamental limitations which I will identify.

Thus the factors limiting the construction and use of electron and proton machines are indeed different. I would not go as far as the recent remark of Carlo Rubbia who said that in the future we have to choose between machines (meaning proton machines) we do not know how to use and machines (meaning electron-positron colliders) which we do not know how to build. One could, however, ask more precisely at what energies do we no longer know how to extricate new events from old ones at proton machines, and at what energies we encounter fundamental technical limits in building electron machines which we do not know how to overcome.

Let me illustrate the proton machine data analysis dilemma with some illustrative figures. The SSC at its rated luminosity of 10^{33} cm^{-2}sec^{-1} will produce about 10^8 total interactions per second. Yet the luminosity is expected to yield fundamentally new events corresponding to the formation of masses in the several TeV range at a rate of perhaps 100 to 1000 per year. Thus the number of "interesting" events is perhaps 1 in 10^{12} of the total hadronic reactions. In addition the recording of a single event requires the writing of perhaps 10^6 to 10^7 bits of information and thus if everything the SSC produces is to be written (somehow), this would imply 10^{15} bits per second. This is unattainable by foreseeable technology and in fact recording events at a rate something like a few per second is more reasonable. These raw numbers, of course, exaggerate the actual situation since many of the background processes occur at such small angles that they escape down the beam pipes and are not observable at any rate. Yet this means that recording has to be "triggered" in such a way that all but perhaps one in 10^8 events are rejected.

The experimental consequences of this circumstance have been studied extensively for processes which have been theoretically conjectured in the new energy range accessible to the SSC. Conceptual detector designs and triggering algorithms which go with them to isolate such processes have been conceived. Although in these terms the problem appears tractable, the tantalizing question remains whether and how one can intelligently develop selection criteria which may not also reject the discovery of phenomena which have not been conjectured in advance to occur. In other words, what is the extent to which we are negating the discovery potential of very high energy proton machines by the necessity of rejecting, *a priori*, the events which we cannot afford to record?

Physicists have carried out extensive calculations which simulate the process of discovering new generations of particles. These simulations take into account the backgrounds produced from known processes, and therefore a "range of discovery" can be defined over which each new collider could have power to observe new objects of certain characteristics in a given range of mass. Not surprisingly the SSC, being the highest energy collider now considered technologically feasible, has the largest range of discovery. However there are holes in its coverage where well-established processes simply cannot be distinguished from new phenomena. The technology of electron-positron colliders which is now available limits the expectancy for discovering masses as heavy as those attainable by the SSC but their range of discovery, as far as it does reach, is more comprehensive and clean, because of the fact that the background goes down as rapidly as the signal.

This leads me to a discussion of the design limitations of electron colliders. Because of the need to increase the luminosity of such machines quadratically with energy it follows that the beam density has to be increased dramatically as the energy is increased. Let me recount some history. In the early days of accelerators, most experiments were done with beams impacting solid targets which might contain perhaps 10^{22} nuclei per cubic centimeter. When the interest shifted to colliding beam machines, in order to use the available energy in the center-of-mass of the colliding particles more efficiently, beam densities dropped back to the order of 10^{13} particles per cubic centimeter because the density of such beams is very much less than that of ordinary matter. To compensate for this loss in target density and therefore interaction rate, detectors for colliding beam machines are generally designed to totally surround the interaction point covering almost the entire 4π solid angle.

Now as the energy of interest moves up even further we need both a density of targets comparable to or even above that of ordinary matter, and detectors which totally surround the collision point in order not to throw particles away. Actually there are additional reasons for requiring what we call a hermetic detector, that is a detector totally surrounding an interaction point. Not only do we have to be greedy in terms of not throwing away any interesting interactions, but our ability to reconstruct any particular event depends critically on being able to detect and track essentially all the charged particles which are produced in a single event. I note that in the TeV region each event might contain tens or even hundreds of charged particles, and the presence of neutral, weakly interacting particles is generally established through an observed momentum imbalance of the visible charged particles. Therefore increasing beam densities and providing "hermetic detectors" are both essential features of machines and detectors for use in high energy physics.

The path to increased beam densities has been initiated in the Stanford Linear Accelerator Center (SLAC) Linear Collider (SLC). Here electrons and positrons are produced, accelerated to 50 GeV, focused, and brought into collision at densities approximating 10^{19} electrons per cubic centimeter. (This is in contrast to the 10^{13} electrons per cc in ordinary storage rings.) This is achieved by controlling the magnetic optics in such a way that the beam diameters are approximately one or two microns and the beam length of the bunch is about 1 mm. The principle of that machine is not scalable to higher energy since it takes electron and positron beams accelerated in a single machine and bends their orbits into collision, as is shown in Fig. 6.

Once the electron energy substantially exceeds that of the SLC, then the synchrotron radiation produced in the bending magnets would cause excessive disturbing effects on the beams. Thus linear electron-positron colliders of the future going beyond the energies attainable at the SLC and LEP must employ an "honest" linear collider principle in which two independent linear accelerators hurl particles at one another.

The question is -- are there any evident limits to that approach? Reluctantly I would deem the answer to that question to be "yes."

There are, of course, economic bounds. If one simply scales the practice of existing linear accelerators to energies into the TeV region, then costs would exceed those of the SSC. Thus technologies yielding economy superior to that of existing linear accelerators would have to be developed.

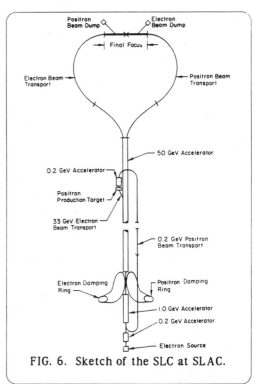

FIG. 6. Sketch of the SLC at SLAC.

Fortunately there are several promising paths in sight which might accomplish this. Beyond simple fiscal economy there is the matter of length. It is clearly unattractive, if not prohibitive, to build a pair of linear accelerators to attain energies of many TeV, that is many thousands of GeV, of electrons and positrons by simply scaling the length of the present 50 GeV linear accelerator at SLAC which is already 3 kilometers. Thus a higher energy gradient must be attained beyond that of 17 MeV per meter now obtained with the SLAC SLC. Again this appears possible; in fact a gradient of 200 MeV per meter is not unreasonable in conventional structures, in which high radiofrequency powers are guided in a disk-loaded waveguide to produce accelerating electric fields.

Many creative physicists have worked to attain extremely high gradients by means other than those used in conventional linear accelerators. High-powered laser beams have intrinsically high electric fields and schemes have been designed to put these high field gradients to work for the production of high energy particle beams. However promising the attainment of high field gradients by some of these schemes may be, I tend to be pessimistic about their applicability to solve the problem of reaching ultra-high energies in practical colliders. The reason is that a collider, to be useful for research in particle physics, must provide both the desired collision energy and generate the required data rate, as discussed above. As we will show, this implies that the average beam power for electron beams required to produce useful interaction rates enters the multi-megawatt range. This result is very general and independent of the mechanism by which these beams are accelerated.

Considering these high beam powers, the efficiency with which electric power from the power line is converted into beam power becomes of paramount importance in the ultra-high energy collider regime. This has not been the case in past accelerators: here the power requirements of the laboratory as a whole tend to be a great deal larger than the relatively moderate power carried by the beam itself. Also, as we will show shortly, an effective electron-positron linear collider demands that the quality of the fields which do the acceleration be very rigidly controlled.

Considering the need for power efficiency as well as precision of field control, it is very unlikely that the various schemes proposed for using laser beams as a means to attain very high energy field gradients can compete in the foreseeable future. This makes the outlook pessimistic that laser and plasma devices, however large their attainable gradients may become, can provide the required parameters. For all these reasons there now exists consensus that the next generation of colliders for electrons and positrons should be based on the "tried and true" technology of disk-loaded waveguides fed by radiofrequency power.

Let me mention a few quantitative considerations. The luminosity L of a collider (essentially the data rate per unit of cross section) is given by:

$$L = N_+ N_- \, f \, / \, A$$

where N_+ and N_- are the number of positrons and electrons per colliding bunch; f is the frequency of these collisions and A is the transverse area of the colliding beams. In turn, this expression for luminosity determines the average beam power through the relation (assuming $N_+ = N_- = N$):

$$P \, / \, L = (2 \, E \, A) \, / \, N$$

In principle this power-to-luminosity ratio could be minimized by reducing the beam cross-sectional area and increasing the number of interacting particles. But there are fundamental limits. The cross-sectional area is limited by the fundamental brightness of the incident beam, to borrow an expression from classical optics. Brightness is the ratio of the square of the beam current to the transverse phase space occupied by the particles; by phase space we mean the product of the beam radius and angular divergence of the beam. According to fundamental thermodynamic principles expressed through what in classical mechanics is called Liouville's Theorem, the radial phase space cannot be modified by optical devices. As a result a practical optical system which produces the final focus in a linear collider cannot decrease the area of interaction below a certain limit.

If the particle is accelerated the quantity known as the invariant emittance, which is the product of the radial phase space times the particle energy, can never be decreased; it can, however, be increased or at least distorted through various imperfections in the accelerating process. As a result in designing a linear collider one has to be concerned with producing a very small invariant emittance for both electrons and positrons and preserving that emittance during the process of acceleration.   Both of these jobs are difficult. The initial emittance for an electron beam depends on the design of the electron source or gun and is limited by fundamental space charge considerations. Positrons have to be produced by conversion of electrons in a thick target, and the scattering and energy dissipation in that target limit the brightness of a positron beam which is practical. It is possible, Liouville's Theorem notwithstanding, to shrink the emittance by so-called damping or cooling de-vices.   An example is a damping ring: here the electromagnetic energy radiated from electron or positron beams trapped in an an-nulus effectively cools the radial motion of the beam and thereby shrinks its emittance.

However, even that process has its limits. If the cooled beam is ac-celerated, then mechanical im-perfections and particularly er-rors of alignment of the acceler-ating structure and focusing ele-ments surrounding that structure can make the emittance grow again.   In order to achieve the type of emittances needed for practical ultra-high energy lin-ear colliders one is talking about alignment tolerances in the mi-cron region which must be main-tained over many kilometers of length. This growth in emittance originates from beam-wall inter-actions and increases with the number of particles N in each bunch.

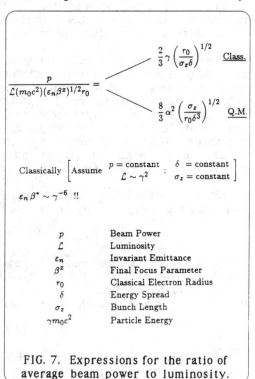

$$\frac{p}{\mathcal{L}(m_0 c^2)(\varepsilon_n \beta^z)^{1/2} r_0} = \begin{cases} \frac{2}{3}\gamma\left(\frac{r_0}{\sigma_z \delta}\right)^{1/2} & \underline{\text{Class.}} \\[2ex] \frac{8}{3}\alpha^2\left(\frac{\sigma_z}{r_0 \delta^3}\right)^{1/2} & \underline{\text{Q.M.}} \end{cases}$$

Classically $\left[\text{Assume} \begin{array}{ll} p = \text{constant} & \delta = \text{constant} \\ \mathcal{L} \sim \gamma^2 & ; \sigma_z = \text{constant} \end{array}\right]$

$\varepsilon_n \beta^* \sim \gamma^{-6}$!!

p	Beam Power
$\mathcal{L}$	Luminosity
ε_n	Invariant Emittance
β^z	Final Focus Parameter
r_0	Classical Electron Radius
δ	Energy Spread
σ_z	Bunch Length
$\gamma m_0 c^2$	Particle Energy

FIG. 7. Expressions for the ratio of average beam power to luminosity.

A further limitation on the number of electrons which can be accelerat-ed in a given bunch is the so-called beamstrahlung effect. This results from the fact that as beams are focused to very small diameters, then the electric and magnetic fields at the edge of the beam become very large and those fields cause the particles in the opposing beams to radiate. For in-stance, in the SLC, the magnetic field at the edge of one beam is about a megagauss. This radiation, in turn, will spread out the otherwise near-mo-noenergetic beams which make their use in experimentation more diffi-cult.

These effects have been extensively analyzed. One can write generalized equations (as in Fig. 7.) which determine that the ratio of average beam power to luminosity is a general function of the invariant emittance prac-tically attainable, the energy spread due to <u>beam</u>strahlung which is tolera-

S L C COMPARED TO 1 TeV LINEAR COLLIDER

	SLC	SLC TECH.	NEW TECH.
ENERGY (GeV)	100	1000	1000
REP RATE (Hz)	180	360	90
LUMINOSITY (CM^{-2} S^{-1})	6×10^{31}	10^{33}	10^{33}
ACCEL GRAD (MV/M)	20	20	200
R.F. FREQ (GHz)	2.86	2.86	11.4
PEAK R.F. POWER PER M. OF ACCEL (MW)	20	20	1200
LENGTH (KM)	3	50	5
WALL PLUG POWER (MW)	50	500	100
$\sigma_X * \sigma_Y$ (μM)	1.6 X 1.6	0.4 X 0.4	$1 \times (5 \times 10^{-3})$

TABLE I. Comparison of design study parameters between the SLC and a 1 TeV linear collider.

ble, the focal length of the final focusing elements which is practical, and the length of each bunch of electrons and positrons which is being brought into collision. As a result of these scaling relationships, the mechanical tolerances which have to be met by the accelerating structure and the final focusing system very rapidly become difficult to meet. Moreover in order to achieve successful collisions in the multi-TeV region one is talking about beams which must be brought into collision at diameters in the angstrom range.

Now, nevertheless people are difficult to discourage, and design studies have been made in Europe, the United States, and the Soviet Union to see how the linear collider principle might be extended to higher energies. I show in Table I the results of some of such design studies. In the first column is the machine collision energy of 100 GeV of the SLC, and the luminosity is 6×10^{31}, which would give an adequate data rate. The length is 3 kilometers, and the beam dimensions are 1.6 by 1.6 microns. If one would simply extend that technology to 1000 GeV, then various things happen. The length becomes very large, 50 km; one talks about 500 megawatts of beam power, clearly impractical; one talks about 0.4 by 0.4 micron beams. The main reason why this extension is so difficult is because the luminosity has to increase to maintain the data rate.

Examples of Beam Parameters for $E_{c.m.}$ of 0.1, 1 and 10 TeV

$E_{c.m.}$ (TeV)	0.1 (SLC)	1		10	
L (cm sec)	2.4×10^{30}	2.4×10^{32}		2.4×10^{34}	
β^*(m)	7.5×10^{-3}	7.5×10^{-3}	10^{-3}	7.5×10^{-3}	10^{-3}
ε_m (m_0c-m)	3×10^{-5}	1.5×10^{-6}	1.5×10^{-6}	10^{-6}	10^{-6}
$\varepsilon_m \beta$	2.25×10^{-7}	1.125×10^{-8}	1.5×10^{-9}	7.5×10^{-11}	10^{-11}
σ_r (μm)	1.5	0.1	0.038	2.7×10^{-3}	1×10^{-3}
N(e)/bunch	5×10^{10}	10^{10}	2.4×10^{9}	2.4×10^{9}	8.8×10^{8}
σ_z (mm)	1	0.25	0.25	10^{-3}	10^{-3}
D	0.63	0.63	1.1	0.09	0.24
H (D)	1.5	1.5	4	1	1
δ	0.8×10^{-3}	0.226	0.3	0.3	0.3
2 P_b (MW)	0.144	3.6	0.760	14.7	5.38
f_r b	180	2300	2000	3800	3800

TABLE II. Comparison of design studies for the SLC (0.1 TeV), a 1 TeV, and a 10 TeV collider.

At an energy of one TeV you can look at a design which is possibly feasible, but there are some extremely critical parameters which certainly would take a great deal of work to explore. If one goes even further than that, if one goes to even higher energy, then the parameters in question become even more difficult.

In Table II I show design parameters of an even more futuristic nature. This is in more detail. In the left column is the SLC, as we know it today. Again, let me note the radius of 1.5 micron for the beam. The middle column incorporates two designs for a machine producing 1 TeV center-of-mass energy. A 10 TeV machine, of two designs, is described in the right hand column. In that case one talks about a 25 angstroms or 10 angstroms radius in the final beam and other parameters which appear to be also extremely difficult. Using "flat," i.e. non-circular beams, may improve this situation significantly. The most important thing is the alignment tolerance which is extremely difficult to meet because the invariant emittance, epsilon, which determines the brightness of the beams has to be improved by three orders of magnitude beyond current practice. So this table which extends our consideration to linear colliders in the 10 TeV range indicates that we are talking about technological achievements that have not been attained. And may I recall the fact that this brightness parameter (the invariant emittance) has to be improved by the 6th power of the energy.

The above considerations which I am outlining here only qualitatively, show when elaborated quantitatively, that using known technology for electron-positron colliders we may just barely be able to reach the TeV range of collision energies but we cannot go much beyond that without some drastic changes in basic assumptions. Let me remind you that none of these deliberations depends on how the acceleration is obtained, whether it is done by conventional radio frequency structures, or whether it's done by laser devices or plasma devices, or anything else that you can imagine. These are very general considerations having to do with beam densities, brightnesses, Liouville's theorem, radiative effects, beam-beam interactions and unitarity.

Quite separate from these considerations it is necessary to develop economical power sources which can feed such accelerators; general scaling relationships indicate that such sources should be microwave devices feeding power in the wavelength range of a few centimeters.

The limits indicated by this discussion may be fundamental (and I am conservative enough to believe that they are fundamental) or they may simply indicate a poverty of ideas. However the bases of these limitations are sufficiently firm that it is doubtful that a single invention akin to the ones which have given rise to the exponential growth portrayed in Fig. 1 will be sufficient.

In Fig. 1, I indicated that this spectacular growth of energy was achieved by a succession of technology. As one technology ran out of steam, a new one took over, new invention made it possible to maintain the overall exponential growth. It is doubtful that this will be possible in the future, simply because so many simultaneous inventions are needed in order to overcome the limits which I've quoted. In order to overcome these limits there have to be inventions in terms of efficient microwave or laser power sources, in efficient transfer of electromagnetic energy to a particle beam, and unprecedented improvements in beam quality both in terms of its production and maintenance of the radial phase volume of the beam.

Of course, it is always possible that the fundamental assumption is incorrect, namely that interaction cross sections will continue to decrease quadratically with energy. However, if that assumption is incorrect, then very fundamental quantum mechanical principles will have lost their validity.

Thus I come back to the current assessment that proton colliding beam devices seem to be feasible as rather straightforward extrapolations of currently existing machines, even to energies well above the 20 TeV per beam of the SSC. However the conduct of experiments with such machines shows formidable problems which already require that for efficient exploitation of the SSC detector technologies and data processing methods be developed which go considerably beyond their current status.

For electron-positron colliders, on the other hand, the situation is reversed: the limitation is that the design of the electron-positron colliders themselves imposes demands on power sources and mechanical precision and alignment to an unprecedented extent, and these demands rapidly escalate with energy.

Obviously it would be foolish to identify these conclusions with ultimate "limits" on the future evolution of man-made high energy particle sources. However, it will take a great deal of ingenuity and invention in many areas of science combined to convert these limits from *limits* into mere *impediments.*
□□

AGNAR PYTTE:

Thank you very much for that talk, Professor Panofsky, "High Energy Colliders and Exploration of Small Distances: What are the Limits?" It occurred to me that one of the accelerators must be the size of Stanford's campus. Fortunately for Professor Panofsky and his friends they have, I believe, 8,000 acres, so were able to build a laboratory two miles long, something that would be rather difficult to do at Case Western Reserve.

I'm very pleased now to introduce the next speaker, Dorothy Michelson Livingston. She grew up in Chicago and now lives in New York. She has served as Honorary Vice Chairman of the Michelson-Morley Centennial Celebration. She is speaking on a subject on which she is quite expert, namely her father. She has written a biography of Albert Michelson, called *Master of Light*. What a particularly appropriate title -- "Master of Light." Dorothy Michelson Livingston.

Reminiscences of My Father

Dorothy Michelson Livingston

Wainscott, New York 11975

Good morning. I'm very happy, and not a little -- forgive me -- proud, to be here this morning and to tell you about memories of my childhood and my father. I think one of the first things I remember was his returning from South America. I believe he was in Peru for a lecture, and he brought back a great big shining solid gold medal. I wrote in my diary, if it were only mine, I would sell it and buy a horse.

He also brought back some hummingbird wings, the feathers of hummingbirds, and big large beetles with iridescent wings. And he explained to us that this was not pigment, it was just the reflection of light on myriads of tiny grooves. And he said, "you know, I'm trying to make a grating that would have as many grooves, but it takes me years and years to make one little grating, and these beetles and hummingbirds just do it automatically."

He was a very nice teacher with us children, and he took pains to give me and my sister violin lessons every morning before he went to the laboratory. This never stuck very hard with me because I didn't turn into a violinist, but I did learn to love music, and that has given me a great deal of pleasure all my life. It was particularly true last night and the night before when we heard these wonderful compositions.

He was very good at teaching us. For instance, we had a place in the country up in Canada and we lived on an island and everything came and went in little boats and we had sail boats. And I can remember my father standing on the end of dock with a megaphone and calling to us, "now ready about," if you're tacking. And when you're coming down before the wind, "now you've got to jibe." And above all, "you must bring the boat up gently and make a smooth landing when you come in, don't crash the dock."

He played tennis. He was a good tennis player, and I think he had a certain stroke which was called a Lawford. I believe it was some kind of a slice that made the ball very difficult to receive. He was a good chess player. Winter evenings he taught me to play chess and I had a lot of tennis lessons which have given me pleasure all my life. He played three cushion billiards and I think he was very good at that because he could predict the angles having been working with angles of light. He was very good at three cushion billiards.

On Saturday mornings sometimes, he would take me to the laboratory, and I remember the visits there. He had a scale that was very, very sensitive. I'd pull out one hair of my head to measure the weight of it and put it on the scale and it just staggered like that; with the weight of one hair.

There was the ruling engine which we couldn't disturb. We could look at it through a window or the door, a glass door. It was there, steadily ruling these beautiful slides that he made.

He told me stories of his life at Annapolis Naval Academy. He was lightweight boxing champion there. He told me that he went to class one day and he had not done his math problem. And he was the first person to be sent to the board. He got up there and he looked at the problem, which he hadn't done, and he figured out a way to solve it. It was different from what was in the book, but he got the right answer. The instructor said, you have been cheating, you've got some sort of pony or trot and you will be up for court-martial. Well, when the day came around that he had to go before the board, they said what explanation have you? He said, well give me another problem. So they did, and he solved the problem in front of them with his new original method and came out with the right answer.

After graduation he went to sea for two years and then he came back and became an instructor in physics. And the first problem was to illustrate for the students the measurement of the speed of light - - Foucault's measurement of the speed of light. Instead of just doing this on the blackboard, he decided to build a live experiment. And he bought a little rotating mirror, a mirror that he used, with his own money, and it was done there, and they've got a big white line at Annapolis, the line where he measured the speed of light. And the answer that he got was far more accurate than anything that had been achieved so far.

With this success he began thinking about other things he might like to do. He read a letter from James Clark Maxwell that came to the Nautical Almanac. It described an idea that it might be possible to measure the speed of the earth through the ether by a method of finding out the speed of light going with the earth's motion, and across the earth's motion. But Maxwell said this is much too difficult a quantity to be measured; it would be impossible. Michelson thought differently, and with some encouragement he decided to go to Berlin for future study.

He studied with the great Hermann von Helmholtz and after a year, this is 1880, he received a grant from Alexander Graham Bell. This gave him the money to have the mirrors built and polished. He set up the first experiment in the basement of the University of Berlin. But there were traffic vibrations, even in those days. The horses and carriages were pounding over his head and shook the thing. So he moved to the basement of the solar observatory in Potsdam (this was 1881). He there set up all his instruments and worked for several months until he got what we all know was the negative answer.

I went to Berlin for the celebration of the hundredth anniversary of this experiment. My young friend Hans Haubold invited me to come there. It was a very interesting discussion then of the significance of this experiment in the light of modern science.

My father was disappointed, as I guess you all know, that he didn't get the accurate measurement that he wanted, because of the negative outcome. So, he returned to America and resigned from the Navy, and was welcomed at Cleveland here as the first Professor of Physics at the Case School of Applied Science, it was called. And there the happy coincidence of his meeting with Edward Morley, which was really one of the more wonderful things of his time here, his friendship with this great chemist.

Morley encouraged him to build the interferometer again and was exceedingly helpful. I think it was Morley who had the idea of mounting the whole thing in quicksilver so it moved around without any difficulty, as smooth as it could be.

Einstein was I think about 40 years younger than my father, and in 1905 when he made his great thesis he was disappointed that he didn't get instant recognition from four people: Ernst Mach, H. A. Lorentz, Poincare and Michelson. None of the four of them came forward in 1905, and said "this is it." Well, years later he came to America when he was widely recognized and feted from one end of the country to the other. He showed a certain coldness at meetings. My father and he would meet at various scientific gatherings and he was, you know, chilly. And of course, Michelson never liked his theory, so that was another point. So they really never spoke then.

Then my father retired from Chicago to California, and he was working up on Mount Wilson when Einstein appeared out there. And in the evening of his life he was very happy to talk with Einstein. My father was failing in health and he was really quite ill at the end and Einstein used to come to the house. I was there when he came. My father would play him a tune on the violin and he would play the violin. Then father'd get out some water colors that he had been making of California, and the whole friendship was really sealed in a beautiful way.

Well I think I've spoken enough. Thank you very much.
(Applause.)

DR. PYTTE:
Thank you very much, Dorothy Michelson Livingston, for those remarks.

□□

KENNETH L. KOWALSKI:

I would like to introduce the Chair for this session, who is a long-time friend of mine and colleague at Case Western Reserve. He is an Institute Professor there and he's done many things on a marvelous variety of topics in theoretical physics, only about half of which I've understood by this time in my career. This is Les Foldy and he's probably best known to you for the FW Transformation, but he's done a surprising number of other things. The next thing he's is going to do is Chair this session.

LESLIE L. FOLDY:

I'll try to be brief in order to allow the speaker maximum time and allow you to have a leisurely lunch and meet your friends. It's my pleasure to introduce the next speaker. Albert A. Michelson is best known for the Michelson-Morley experiment, his measurements of the velocity of light, his determination of the standard meter in terms of the wave length of an atomic emission line, and for his stellar interferometer which made possible measurements of the diameters of neighboring stars. It is less well known that he's also made several important discoveries in atomic spectroscopy. The first of these was a hyperfine structure of spectral lines, an effect now known to be associated with the structure of the atomic nucleus. This was some 20 years before Rutherford discovered that atoms had nuclei.

The second was a fine structure in spectral lines of hydrogen, now known to arise from a relativistic effect in the orbital and spin dynamics of that atom. Some 35 years later, this hydrogen fine structure was to play an interesting role in the relativization of quantum mechanical Schrodinger equation for the hydrogen atom. Thus, it would be quite proper to say that Michelson was a progenitor not only of special relativity and classical physics, but quantum physics as well. Therefore, it is particularly fitting, that the opening session of this symposium should include a talk by a physicist who's been honored with a Nobel Prize in Physics for his contributions to the development of laser spectroscopy, including its applications for atomic spectroscopy.

Professor Arthur L. Schawlow, though born in the United States, took his Bachelors, Masters and Doctorate degrees from the University of Toronto, which later, among other universities, further recognized his contributions with an honorary doctorate. He's currently J. G. Jackson - C. J. Wood Professor of Physics at Stanford University, whose faculty he joined in 1961 after spending ten years at the Bell Telephone Laboratories. In 1958 he wrote a seminal paper with Charles Townes outlining the conditions for producing an optical laser. A few years later both the ruby laser and the gaseous laser became realities. He's been frequently honored for his work in optical and microwave spectroscopy, lasers and quantum electronics, culminating, as I mentioned earlier, by the award of a Nobel Prize in 1981. It is my pleasure to present Professor Schawlow, whose subject will be "Atoms, Molecules and Light".

Atoms, Molecules and Light

Arthur L. Schawlow

STANFORD UNIVERSITY

Stanford, California 94305

That's a big subject and obviously we can't cover all of it, and so we'll just select a few things. When I was President of the Optical Society of America in 1975, we appointed a new editor to the Journal. In his first editorial, he said, "The physics of light is by now pretty well understood, and the real interest is in applying this knowledge for useful purposes: i.e. engineering" (1).

I never did agree with that because it seems to me that light is very complex. We know its properties only through its interaction with matter, and those interactions are as complicated as matter itself, and as diverse. For instance, they depend on the wave length of the light, and you can get very different effects with different materials. Some phenomena depend dramatically on the intensity of the light, and so on. However, there still are some problems in optical engineering, such as the design of a DASAR. This mythical device, whose name is an acronym for Darkness Amplification by Stimulated Absorption of Radiation, would be a sort of black searchlight, projecting a beam of darkness instead of light. It is a concept that cartoonists have illustrated (2), but their drawings leave its principle of operation as, I presume, an exercise for the reader.

However, we do have a useful family of devices, lasers. Lasers have been used and are being used to learn more about the fundamental nature of atoms, molecules and light. We start with the properties of laser light. It is directional, powerful, monochromatic and coherent. From these arise its uses. Of course now, just as all men are equal, but some are more equal than others: so all lasers have these properties. But some lasers produce output powers in the microwatt range, and some in the megawatt. So their powers are quite different.

Lasers and spectroscopy are very closely related. I did a search through the Lockheed Dialog INSPEC file for papers which mention lasers and also something to do with spectra in their abstracts. Figure 1 shows that the number of papers involving lasers has risen quite sharply from 1969 to 1983, and that about a quarter of the papers in that field also mention spectra. That is partly because lasers are used to study spectra, partly because spectra are the raw materials of lasers. At any rate, they are very closely related, and both growing very rapidly.

In Fig. 2 is a spectrum that everybody should see at least once in his life. It is difficult to photograph because it is hard to find color film that has

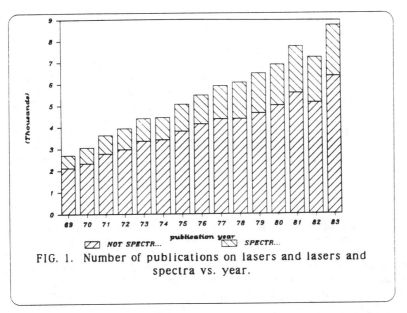

FIG. 1. Number of publications on lasers and lasers and
spectra vs. year.

any sensitivity in the ultraviolet. This is the spectrum of atomic hydrogen,
the Balmer spectrum, for which Balmer gave an equation back in 1885.

FIG. 2. The hydrogen Balmer spectrum. (See Schawlow
page in color insert.)

The red line in this spectrum is H-alpha, the one for which Michelson
discovered the fine structure splitting and which was so important in leading to Dirac's relativistic theory of quantum mechanics and then, later, to
the discovery of Lamb and Retherford, who actually measured that fine
structure. That measurement in turn was an important stimulus for quantum electrodynamics, developed through the work of the theorists
Feynman, Schwinger and Tomanaga.

Balmer's work was pure numerology. However, as you look at this spectrum, it looks so simple that you really feel that it could be understood, and
indeed it does fit a rather simple equation.

$$1/\lambda = \nu = R/n^2 - R/2^2$$

This is Balmer's equation for the wavelength in modern language; a very
simple thing. Here R is the Rydberg constant, named after the Swedish scientist who developed a similar formula for the alkalies a few years after
Balmer's theory. Interestingly enough, Rydberg didn't know about
Balmer's equation, and was glad to see it was a special case of his. Later,
in 1912 when Niels Bohr was working on the theory of the atom with
Rutherford's group in Manchester, he didn't know about Balmer's formula
either. Bohr told T. D. Lee and Abraham Pais years later that he had learned
about it from a chemist friend named Hansen shortly before he was to leave
Rutherford's lab. As soon as he saw this formula he was able to work it out

from quantum principles quite quickly.

Now, I wondered how could it be that a physicist didn't know this formula. Certainly they all do nowadays, or at least they've heard of it. But I asked Lee how that was possible, and he said, "Oh, we asked him that, and he said in those days spectra were considered something very complicated, like the notes on a piano. They depended on some complicated details of the atomic structure, and they were not a serious part of fundamental physics." It is interesting that up to about 1890, the 19th century physicists really tried to fit spectra as normal modes of some kind of oscillators. But that didn't work out, although there were those tantalizingly simple formulas that indicated it might work out somehow. But then in the 1890's and 1900's the attention shifted to the particle nature of matter, the electron and then the nucleus were discovered, and people just forgot all about fitting spectra.

And too, I think Einstein once said, that he had not known specifically about Michelson's experiment. Well, it was 20 years or so earlier, but it was part of the general culture of physics in those days, because others like Lorentz and Poincare and Fitzgerald had thought hard about it. And I'm sure that Einstein knew some of those other things, even if he didn't know the experiment directly. So, things do get forgotten and rediscovered.

And for you students, let me say, if you ever run out of ideas, just go the library, pick out an old issue of *Physical Review*, about 20, 30, 40 years old, and look through it. You'll find all sorts of unfinished business. It's like being the Connecticut Yankee in King Arthur's Court. You can go back with modern knowledge and fill in, go on from there, see things that haven't been done. It's a rich source if you ever need some fresh ideas.

FIG. 3. Laser light is monochromatic. (See Schawlow
page of color insert.)

Figure 3 shows that lasers can be monochromatic. It is a picture of a helium-neon laser with a bare glass tube. It is photographed through a diffraction grating, a holographic diffraction grating, incidently. You can see the spontaneous emission from the side of the tube has many, many wavelengths ranging throughout the visible. The laser beam, which is seen here by letting it graze along the screen, has only one color: it is monochromatic. Now this shows only to one percent, but it can be made monochromatic to a part in a hundred million or better, if you're willing to work at it.

However, the early lasers were not tuneable and it wasn't until we got dye lasers through the work of Sorokin and Lankard in this country and Schaefer in Germany that we were able to make tuneable lasers that could extend over a substantial range in the visible.

FIG. 4. Simple dye laser. (See Schawlow page of color insert.)

Figure 4 shows a simple, tuneable dye laser consisting of a dye cell containing some dye such as Rhodamine 6G. Many, many dyes that are fluorescent will work for such lasers. The output mirror is on the left. The other mirror in this case is a diffraction grating, which is a good mirror for one wavelength at a time. As you turn the micrometer screw you resonate the system to different wavelengths and therefore have an output at different wavelengths. You can get dye lasers at any wavelength in the

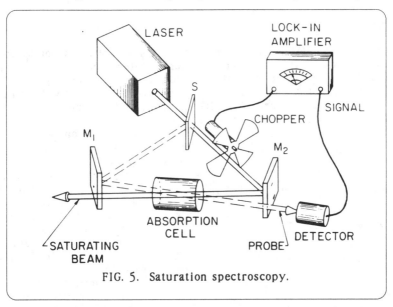

FIG. 5. Saturation spectroscopy.

visible or near-visible regions. Other tuneable lasers exist for some other regions. Of course, they're more complex than this and have gotten more and more expensive over the years as people work to make narrower lines from them. This one would be too crude to do very much spectroscopy.

A big advance came in 1970, when Hansch at Stanford, and Borde in France, independently developed a scheme for using the brightness of the laser and directionality as well as its monochromaticity to eliminate the Doppler broadening which always plagued the spectroscopist.

Now, I'm sure that all of you that are physicists know about the Doppler effect. You used to be able to teach your classes about it by telling them about train whistles whose pitch changes as the train goes by, but they don't seem to have many trains anymore. So, let me just show you how to tell them. You see, as I move toward you my voice goes up in pitch, (demonstrates); as I move away it goes down in pitch. That is the Doppler effect... slightly exaggerated. So, if you have atoms in the gas that are free and not bound, then they will be moving around rapidly with thermal agitation. Therefore, the light from some atoms will be raised up in frequency, while from others it will be lowered, and so the spectral lines will be broadened. This may wipe out fine details of the spectrum, as might result from hyperfine structure.

Now, you can get around that by making an atomic beam and observing in a direction perpendicular to the atomic velocities, and that's what I did for my graduate thesis. It's was difficult because we had to excite the atoms in the beam with electrons. Now it is easier because you can excite them with lasers.

Figure 5 shows Hansch and Borde's method of saturation spectroscopy (3-4). It makes use of the fact that the laser is intense enough that it can at least partly bleach, i.e. reduce the absorption of the absorbing gas. Then a probe beam from the same laser going through in the opposite direction will be able to pass through. Actually these beams are almost exactly opposite, not at as big an angle as shown here. As the saturating beam is chopped on and off, the probe beam then is modulated. But, that happens only if the two beams can interact with the same atoms. So, the laser must be tuned to those atoms that are standing still, or at worst, moving transversely, because atoms that are moving either to the right or to the left see one beam Doppler shifted up in frequency, the other one Doppler shifted down. The moving atoms can't be resonant to both beams at the same time. So, this method picks' out those atoms that happen to be standing still. Hansch soon applied this to look at the spectrum of hydrogen, which was really the simplest one, and would interest the theorists most (5).

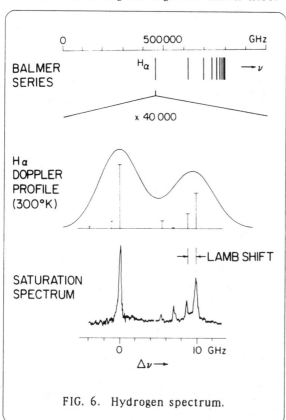

FIG. 6.   Hydrogen spectrum.

Figure 6 shows the results obtained by Hansch and Shahin with the saturation method on hydrogen. At the top is the Balmer series and next is the H alpha line expanded 40,000 times. A perfect spectrograph at room temperature would show no more than that profile. But from radio frequency studies, such as those of Lamb and Retherford (6) it was known that those fine structure components shown lay beneath the line. It had not been possible to resolve it any better, because hydrogen is very light, and the atoms move around very fast, so that the Doppler broadening is especially large.

However, in this earliest experiment of Hansch and Shahin, using the saturation method and a rather crude pulsed laser, they were able to resolve the fine structure well, as shown at the bottom of Fig. 6. The Lamb shift was thus seen optically in hydrogen for the first time. You might think the sensible thing to do would be to measure these spacings, once that they are resolved, but they had indeed all been measured very well by radio frequency methods. Instead, Hansch and his associates concentrated for the next few years on measuring accurately the absolute wavelength of one of these components, in order to determine the Rydberg constant, which measures the binding of the electron to the nucleus.

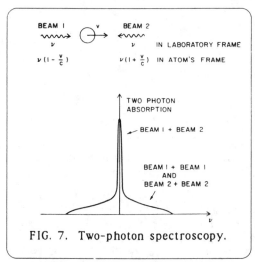

FIG. 7. Two-photon spectroscopy.

Figure 7 shows another method of getting rid of Doppler broadening, when you have a two-photon absorption, or two-beam absorption. This is again a nonlinear effect where the laser is intense enough that it modifies the properties of the atom, so that it is capable of absorbing light from two beams at the same time. This gives a transition between states of the same parity, whereas the ordinary, single-photon absorption is between states of opposite parity. This method is one of the things that is so simple that I feel stupid not to have thought of it. But it was proposed instead by Vasilenko, Chebotayev and Shishaev in Russia (7) and brought to this country by Letokhov about three years later because it seems no one here had read their paper.

The method is this: atoms are exposed to light of two beams in opposite directions from two beams from a laser tuned so that two of its photons can produce the transition between the two states of the same parity. If the atom is moving with a velocity v along the beam direction, one beam will appear Doppler-shifted up in frequency to $v(1+ v/c)$, and the other one will be shifted down in frequency to $v(1-v/c)$. However, the sum of their frequencies, and therefore of their photon energies, is independent of velocity. That is, if the atom absorbs one photon from each beam, the $+v/c$ and the $-v/c$ just cancel and the sum of their frequencies is just $2v$, as needed for the two-photon transition. Thus all of the atoms, and not only those that are standing still, can contribute to a very sharp two-photon absorption. There is a weak broad background, but usually too weak to see, from atoms which absorb two photons from the same beam, so that you just see this very sharp two-photon absorption. It's a very simple thing to do, once you know about it.

Both the one- and two-photon methods have been applied to study lines in the spectrum of atomic hydrogen. The energy levels are shown in Fig. 8. Hansch applied the one-photon saturation method to the H alpha transition, n = 2 to 3. However, even when Doppler broadening is overcome, most spectral lines are broadened because most of the excited states have short lifetimes, decaying rapidly by radiation to lower states. The sole exception is the 2S state, which is truly metastable and has a lifetime around 1/7 second. It can't decay except by a very weak two-photon emission. So, you can get very sharp lines from the 2S state, although most lines are broadened by rapid radiative decay of the upper state.

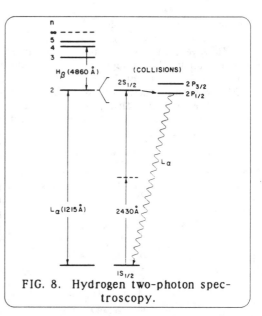

FIG. 8. Hydrogen two-photon spectroscopy.

States with higher n values do h ave some what longer lifetimes. Zhao, Lichten, Layer, and Bergquist (8) have measured the 2S to 3P and 2S to 4P transitions in an atomic beam. Biraben, Garreau and Julien (9) have measured the transitions form 2S to 8D and to 10D. On the other hand, if you go down to the 1S to 2S transition, which can be observed by Doppler-free two-photon absorption, the resonance can be a very sharp line indeed. Ultimately, the width imposed by the lifetime on this line would be about one Hertz out of 10^{15} or so. If you could meure to one percent of the linewidth, that would be an accuracy of one part in 10^{17} !. But nobody measures anything to that accuracy! So, I'm sure the scientists are going to be working for a long time on this 1 to 2 transition.

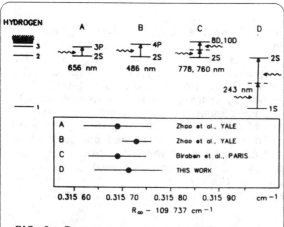

FIG. 9. Recent measurements of Rydberg's constant.

Any one of these lines could give a precise value of the Rydberg constant, if the atomic theory is adequate. According to the Bohr theory, the energy difference between states with n = 1 and 2 should be just 4 times the energy difference between states with n = 2 and 4. If the ratio is not exactly 4, it is because there is a Lamb shift in the lowest, 1S, state. So one can measure the Lamb shift of this 1S state. There is no nearby level like the 2P to compare it with, so that this is the only way you can measure the Lamb shift of the ground state (10).

Figure 9 shows the results of some of the experiments that have been done. All of the measurements are in pretty good agreement. Indeed, the Rydberg constant, instead of being known to a part in 10^7 as it was a few years ago, is known to about 3 parts in 10^{10}, and obviously you can go farther.

What else can you do? You can test quantum electrodynamics, but you run into two problems here. One is that the structure of the nucleus is complex and the other is that quantum electrodynamics gets awfully complicated when really precise results must be calculated.

I heard Dirac talk a few years ago and he said he didn't like quantum electrodynamics (11). He thought it an ugly theory, not beautiful like Heisenberg's quantum mechanics, because it's a very complicated calculation. You have to use hundreds of diagrams. If the pre-Copernican astromomers had had modern computers they could have added a couple of hundred more epicycles and obtained a very good account for the motion of the planets without having to bother about this heliocentric nonsense!

But of course, to be fair to quantum electrodynamics, it has stood every test so far. It has a systematic prescription for determining these epi......, I mean not epicycles.., but these diagrams that they have to use. Now there are also beautiful measurements made on positronium by Chu and Mills at Bell Labs (12), a very difficult experiment. They have gotten good results with the assistance of John Hall from the Joint Institute for Laboratory Astrophysics. (2) There seems to be a small discrepancy between the the prediction of theory and the experiment, so that more measurement time and improved theoretical calculations will be needed before they can be sure whether the discrepancy is real. Positronium is somewhat complicated to calculate because both particles have small mass, and there is an annihilation term that effects the energy.

Mills and Chu have also just recently observed the spectrum of the same line in muonium, which is a really elusive atom consisting of an electron and a short-lived positive muon (13). (They had to go to an accelerator in Japan and worked for quite a long time too to get a few counts.) The results so far are pretty much in agreement with quantum electrodynamics.

The precision of the measurements on these various simple atoms can certainly be pushed at least another factor of a hundred, and they will be because the challenge is there. It gets more complicated and expensive. It appears that we are almost at the limit of the accuracy that one can reach by measuring wavelengths and converting to frequency, at this level of three parts in 10^{10}. It may be that to go farther we will have to make direct frequency measurements in the optical region, which is difficult and complicated, but not impossible. Maybe somebody will get clever about that. The challenge to test quantum electrodynamics better and better is clearly there.

Another thing that Hansch has mentioned: you measure these frequencies in terms of the standard frequency, which is the cesium clock (14). But there's nothing very fundamental physically about cesium. What may be more interesting is to make really precise measurements of the ratios of the frequencies of various lines within the hydrogen spectrum. By so doing you can eliminate various parameters from different combinations of these transitions (14). Further information can be obtained from the heavy hydrogen isotopes, deuterium and tritium.

When I was a graduate student at the University of Toronto, I did atomic physics under Professor M. F. Crawford. I saw a lot of people around us working on molecular spectra, and I was led to develop

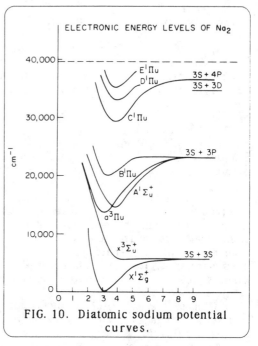

FIG. 10. Diatomic sodium potential curves.

the definition of a diatomic molecule which is: a molecule with one atom too many. Because when you get that second atom you have not only more electronic states than you have in the atom, but you have vibrational and rotational levels, too. You may have perhaps fifty or a hundred vibrational levels for every electronic level, and you can have hundreds of rotational levels for each vibrational level.

Instead of the few lines in the spectrum of the sodium atom, the Na_2 molecule has thousands of lines in the visible and more in the ultraviolet. The electronic levels are customarily shown as in Fig. 10. These are all the levels that were known from fifty years of conventional spectroscopy. The ground state is of even parity, the others are all odd parity states. They have a potential well which shows the binding of the two sodium atoms to each other. When they dissociate, they go to atomic states such as 3s, 3p and so on. So these six excited states were all that were known until we decided that we would try to find a systematic spectrum simplification method using laser level labeling.

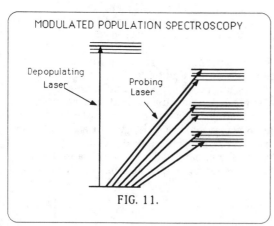

FIG. 11.

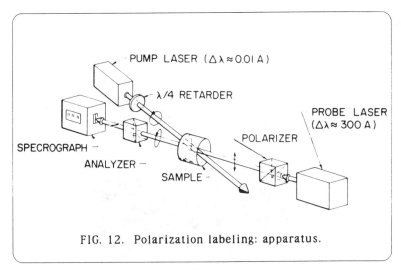

FIG. 12. Polarization labeling: apparatus.

Figure 11 gives an indication of how that works. You take for instance a molecule which has all these levels. You tune a laser to excite one transition from one level up to another level, just one. When you do that, you can depopulate the chosen lower level, or you can polarize it by taking out molecules of a certain preferred orientation. Then you can pick out just those molecules which make transitions from this labeled state. To do this, you can chop the depopulating beam off and on, alternately reducing and restoring the population or polarization of the labeled level (15). Then lines from the labeled level will stand out clearly, because these are modulated. In a diatomic molecule, there will usually be only two of them for each vibrational state, sometimes three. For the allowed lines, the rotational quantum number changes by plus or minus one or in some cases zero. So you can rather quickly pick out these and get a whole series of vibrational levels. Well, we have looked at sodium as a test case and we've gone a little farther in using two lasers to make two steps up: first to put atoms in the A electronic state and then a broadband laser to absorb up to still higher states. So, we go from states of even parity to odd to even again and can rather easily pick out the lines that belong together (16-17).

Figure 12 shows the apparatus. It has a pump laser, which is a pulsed dye laser, which puts atoms up into the intermediate state and orients them. The probe beam has a broad band of wavelengths, but it goes through crossed polarizers. Only at wavelengths of transitions from one of these oriented levels, where the atoms appear asymmetric, can light pass through the analyzer to the spectrograph.

Figure 13(a) shows a few angstroms of the visible spectrum of Na_2; this is the forest of lines you get by ordinary spectroscopic methods, and in (b) we see the result of the polarization labeling method from the ground state to the first excited state of the Na molecule. You see nice doublets: ΔJ is +1 or -1. The vibrational quantum number can be obtained by just counting from the longest wavelength, which corresponds to the upper state vibrational quantum number $v'' = 0$. The missing lines are weak because of the

Franck-Condon factor, which varies for the different vibrational quantum values.

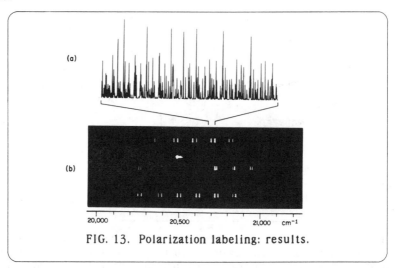

FIG. 13. Polarization labeling: results.

Some results of the labeling technique are shown in Fig. 14. The Na$_2$ molecule has been explored by several of our graduate students. In about five years, they have found 24 more new excited electronic levels to add to the six known previously. The new levels fit into several families of molecular Rydberg states. They are all even parity states and are characterized by angular momentum labels, Σ, Π or Δ. The several series of levels are labeled by their principal quantum numbers and for each of these, the binding

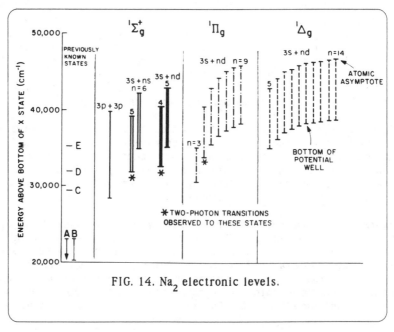

FIG. 14. Na$_2$ electronic levels.

energy, rotational and vibrational constants are measured. It is then possible to extrapolate these to get the constants of the ground state of the Na_2^+ ion. Others have gone on and used a number of other laser techniques to explore molecular spectra. Thus the process of analyzing molecular spectra is becoming subject to systematic techniques.

Why would one want to do that? Well, because it's there. But if you look at the long range view, the task of physics is not just to explain the hydrogen atom, or the nucleus, but all of the natural world. I was involved in a panel discussion a few years back with Roger Sperry, the biologist, who had won a Nobel Prize for his work on the brain. He insisted he didn't think quantum mechanics was going to be enough to explain life. On that panel there were three physicists and a couple of chemists, and we all disagreed with him. We all thought that quantum mechanics, since it covered the energy range that is involved for life, was going to be enough. But, we don't really know. We have to confront the increasing complexities.

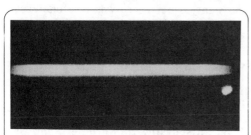

FIG. 15. Sodium beam at room temperature. [Wm. Fairbank, Jr.]

Joining atoms to make diatomic molecules is only the first stage, and elsewhere in this meeting there will be discussion of some of the enormously more complicated actual biological molecules. But I think we do have to try to link their properties as much as we can to fundamental physics, and do these intermediate steps. For many of these, and not just the simplest, lasers are providing new tools and techniques.

In the course of our work on laser spectroscopy, we built a continuous wave dye laser, when those things were new. William Fairbank Jr. tuned this laser to the orange-yellow wavelength that sodium atoms absorb, emit, and scatter. It is shown in Fig. 15: a beam is passing through a cell of sodium at room temperature. There are not very many sodium atoms in the vapor at room temperature; about a hundred thousand per cubic centimeter, and probably only a few hundred in the beam at any one time. And yet here it is very clearly seen. He was able, with photoelectric detection, to observe the spectrum of the scattered light from sodium at temperatures as low as -30°C, where there are only one or two atoms in the beam at any one time. This led us to realize that you really could see single atoms and that the sensitivity of optical spectroscopy should be higher even than with nuclear methods, because you don't destroy the atom when you scatter light off it. It's still there; you can scatter some more light until you have enough to get a good record (18).

Well, it's better of course, if you can make the atom stay there, instead of drifting in and out of the region illuminated by the laser. Neuhauser, Hohenstatt, Toschek and Dehmelt, at the University of Heidelberg, built a little trap which could trap charged barium ions. (19) I won't explain how it works, but I'll show you the results.

In Fig. 16 you will be able to see individual trapped ions. In the upper left there are three barium ions in that one little spot. And in the next one there are two. And in the last one, there is just one single barium ion. And you can't have any less than that. And it's kind of stimulating to think what it would be like if in a few years people can detect a single atom or molecule of any substance. What do you want to bet there won't be one carcinogenic molecule in a quart of milk? There will be, but you'll have to learn to live with limits. But, on the other hand, we might discover new substances that are important for life, but are present in such small amounts they haven't been detected yet.

When you are doing high-resolution laser spectroscopy with atomic beams or with the saturation method, you are still limited by the second-order Doppler effect. For hydrogen, that's going to be a serious limitation eventually. So, in 1974, Hansch and I were discussing this problem, and we realized that you could cool atoms with laser light. The way it's done is illustrated in Fig. 17. A laser is tuned to just below the resonant frequency of the atom, on the long wavelength shoulder of the Doppler profile. Then if the atom is standing still it won't interact with the light, as it will be out of resonance. But if the atoms are moving and you bring in laser beams from all of the six principal directions, then moving atom will still be out of resonance with the beams transverse to it, and so their effects can be ignored. A beam traveling in the same direction as the atom will be even farther out

of resonance, because it will appear to the moving atom to be down-shifted in frequency (20). But the beam coming from the direction opposite to the atom's velocity will be shifted up into resonance. Every time the atom scatters it, there will be a little momentum transferred to the atom, and the atom's velocity will be reduced. This is because the photons have momentum, or you can think of it as classical radiation pressure. The atom loses about a centimeter or two per second every time it scatters one of these photons. And it can do about 10^8 times per second. So the atom can be rather quickly cooled down to a temperature well below one degree Kelvin.

We didn't try laser cooling because we were really interested only in hydrogen, for which there was then no suitable laser.

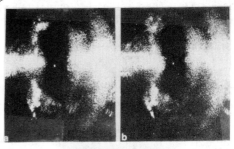

FIG. 16. Trapped barium ions. [Photo: Dehmelt et al., University of Heidelberg. *Spectral line shapes*, de Gruyter, New York, 1981.]

But a few years later, Steven Chu came upon
the same idea.   He mentioned it to Arthur
Ashkin, who has done a lot of work with light
pressure. Ashkin said, "that's a good idea, why
don't you try it?" Chu did it with sodium atoms,
and it worked beautifully (21).

After that Ashkin said, "you know, this isn't
new, Hansch and Schawlow suggested this."
But Chu did make it work. And he did one clever thing. Hansch and I realized that if we were
to cool ordinary room temperature sodium, the
atoms would travel about a meter in the second
or so it would take to cool it. That would be too
long. We would have to have a huge box filled
with laser beams. Chu evaporated some sodium

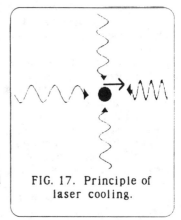

FIG. 17. Principle of
laser cooling.

with a pulsed laser and then waited briefly until the fast atoms were gone.
He then cooled the slowest ones down still slower. In doing that, he had to
tune the lasers to keep in resonance with the atoms as they became slower,
but he was able to do that in a reasonably small volume, by starting with
some moderately slow ones (22).

FIG. 18. Optical molasses. (see Schawlow page in color insert.)

What he obtained was what he called "optical molasses." You can see in
Fig. 18 a little blob of sodium which, if it tries to move anywhere is re-
strained by the viscous force of the light, and nothing else. And he could
hold it there for a fraction of a second. Then it was realized that once the
atoms are that slow they can be trapped and held for a long time. Atoms
can behave as small pieces of dielectric material. An intense off-resonant
light beam, producing a strong optical electric field, will pull the atoms
into the strongest part of the field, just as iron is pulled into a magnet. This
force is very small and therefore nearly negligible for fast-moving room
temperature atoms. But if
you start with these cold
atoms, they can be trapped
at the region of the focus,
where the field is strongest.

Figure 19 shows the ar-
rangement for trapping.
The cooling beams are
shown as arrows, and the
big arrows are the trap
beams focused to give a
high intensity at one place
(22).

In Fig. 20 the trapped
atoms are the little bright
spot at the center of the op-

Atoms
in
molasses

Focused
trap
beam

Atoms
in
trap

Atom trap. *Schematic diagram shows the
six molasses laser beams and the focused
trapping laser beam.*

FIG. 19 Cooling and trapping.

tical molasses. The work of Chu *et al.* on trapping has continued and they have recently demonstrated a still better trap involving a magnetic field, and in it kept atoms for a couple of seconds (23).

Now you can move that atom around once you got it in the trap. It's like optical tweezers; you can move the trapping beam and move the atom around. More recently Ashkin has applied the same technique to trap bacteria with an argon or a Nd:YAG laser under water. He has actually been able to hold them there and watch them multiply. It apparently doesn't bother them too much to have this intense infrared light on them as long as they're cooled in water (24).

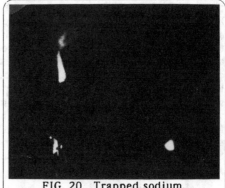

FIG. 20. Trapped sodium.

Speaking of trapping, last year we were all pretty excited to read that Gabrielse and his associates have been able to trap antiprotons (25). At CERN a beam of a antiprotons is slowed down by a degrader. Some of the slow ones are down to a few kilovolts. Then, as indicated in Fig. 21, the antiprotons can be trapped with a magnetic field that prevents them from wandering sidewise, and an electric field, as an endcap, that keeps them from going forward. The antiprotons can only oscillate back and forth. They have been held for 10 seconds or so. With a better vacuum, one could hold them almost indefinitely. Gabrielse is going to do a precision measurement of the mass of the antiproton to see if it's really exactly the same as the proton, which is important to know.

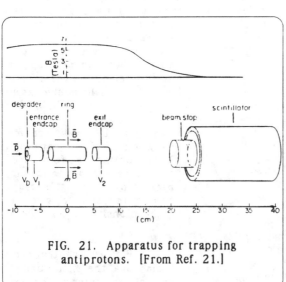

FIG. 21. Apparatus for trapping antiprotons. [From Ref. 21.]

But, it would be nice to be able to keep more of them for longer times. If you have a lot of charged particles, the space charges will blow them apart. What you really need to do is trap these antiprotons, bring in some positrons and make antihydrogen. We already know how to trap atoms. So you could conceive then of having a gram or so of antihydrogen. This could even be floating in a satellite out in space. As long as it doesn't touch anything and is in a good vacuum, it will stay there. And then if you want to release the energy, say for a long space voyage, you just let it touch some ordinary matter. You would get a million, million joules from the annihilation of one gram of antimatter. Now, that's all science fiction, but the pieces are all there. Trap the antiprotons, make antihydrogen, and then trap the atoms. It probably will happen some time. The only question is do atoms of antimatter fall up or down? Can you orbit them or not? That's not really known, despite the strenuous efforts of my colleague, William Fairbank. His experiments, to seek the gravitational force on free positrons, are made nearly impossible by the enormously larger electric forces.

Now, let me turn to another point, and that is the test of quantum mechanics through Bell's inequality. I'm told that Niels Bohr once said that if quantum mechanics doesn't frighten you a little, you haven't really understood it. The predictions of quantum mechanics are familiar. Most of you know how to use it, and it works always, nothing wrong with it. But still when you really realize what it predicts, it bothers people.

For example, Einstein, Podolsky, and Rosen proposed an experiment where you have two correlated photons (they were talking about electrons) that are of the same polarization, but travel in opposite directions. This is illustrated in Fig. 22. As you measure the polarization of one, it will tell you something about the other one. Classically, we would think of this as sending out a polarized wave. Then if it weren't quantum mechanical, you would get a response in one detector that was proportional to the cosine squared of the angle between the polarization direction of the wave and the allowed polarization of the polarizer. The other detector would also have a response proportional to the cosine squared of the angle between the wave polarization and the analyzer setting. Thus if the detectors are fast enough and sensitive enough to detect the pair of photons from a single atomic cascade, classical reasoning would lead us to think that the probability of detecting a photon at each detector would be proportional to the square of the cosine of the angle between the wave's polarization and the polarizer setting. But what quantum mechanics predicts is that the probability of the two detectors responding to a single atomic emission will depend only on the angle between the polarization of these two detectors, as if they were tied together at the same place. And so, the probability of detecting photons at both detectors would go as the cosine squared of the angle between their settings. That means of course there could never be any simultaneous counts if the detectors were perpendicular, because $\cos^2$ of 90 degrees is zero.

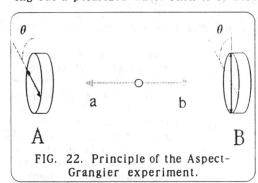

FIG. 22. Principle of the Aspect-Grangier experiment.

But classically, if the beams were polarized at 45 degrees to the two polarizers, you could get some light through each of them. Now this seemed very strange to Einstein: that it really wouldn't depend on what was going on in between the source and the polarizers. J. S. Bell showed that no kind of real signals sent from the source to the detector could reproduce the correlation predicted by quantum mechanics (26). The experiment was a difficult one, but it was done by Clauser and Freedman (27) and subsequently confirmed by others. And more recently, it was done very well by Aspect, Dalibard, and Roger in France (28). They used essentially the arrangement of Fig. 22. The polarizations detected at the two ends are found to be more closely correlated than you

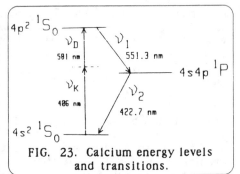

FIG. 23. Calcium energy levels and transitions.

could get for any kind of polarization of waves between the source and the detectors. J. S. Bell had shown that, particularly at angles around 22 degrees between the polarizer settings, you would get more correlation of simultaneous detected photons with quantum mechanics than with any semi-classical picture, or any local realistic picture of anything that could be traveling from the source to the detectors.

To get these two photons polarized the same way, Aspect, Dalibard, and Roger used two lasers to make a two-step excitation indicated in Fig. 23. You go from a state with no angular momentum and no preferred orientation to another one of the same, and then two photons are emitted, one after the other, and they would go sometimes in opposite directions, and they would

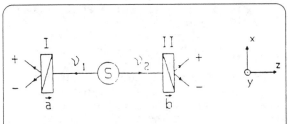

FIG. 24. Aspect and Grangier experiment. [From Alain Aspect, *Atomic Physics* 8, Plenum 1983.]

Figure 24 shows the schematic diagram of the experiment. They used two polarizers, or a polarizing prism, that could measure the component of polarization either in the X or Z planes. They found that the results agreed perfectly with quantum mechanics, with good precision, and this rules out any kind of realistic picture of the two polarizer-detector combinations making separate measurements on a wave existing independently of them. In addition they did an experiment suggested by David Bohm some years ago: that the polarization settings be rapidly changed during the time that light is traveling from the source to the detectors. That is, after the photon is launched, the polarizer settings are changed rapidly, and at least pseudo-randomly. Would there still be the same correlation? The answer is yes, there is. The detectors act as if each one knows perfectly well where the other one is, even though they're forty nanoseconds (forty feet) apart. (The English unit is very useful, one foot equals one nanosecond's travel, at the velocity of light.)

It took 40 nanoseconds for light to go from one detector to the other, and the polarizer setting changed every 10 nanoseconds. So, there's no way that an optical signal could be sent from one to the other. Now, this has really, I think, caused a reopening of the interpretation of quantum mechanics, and I think it's now fairly respectable to wonder if there couldn't be at least, better metaphors for what quantum mechanics does, better than just saying, "Well there's a great smoky dragon there, and all we know is the beginning and the end, when we make measurements." There is a very nice little book, called *The Ghost in the Atom"* by Davies and Bowen, a little paperback book (29). It discusses in a series of interviews with various proponents of the different theories that might account for this strange phenomenon, without just saying, "well, it happens".

And there is also a recent resource letter in the *American Journal* of *Physics* by Ballantine and Jarett (30) and there was an article in *Reviews* of *Modern Physics* last year by Cramer (31). All of these give very good critiques of all the other explanations, all the other views of quantum mechanics. I'm not very happy with what each of these ends up with, but I think it's still an interesting question.

Let me say, on that experiment the interesting thing (as an experimentalist, not worrying about theory at all) would be to move the detectors farther apart than 40 feet. Are they really correlated at a mile apart? Well, we could put SLAC to that use sometime when they finish using it as an accelerator. You've got two miles there. You would need some way of getting more photons going the right direction with random but correlated polarizations. It has been shown recently by Carrol Alley that you can use such correlated photons from a parametric amplifier (32). Thus it is possible, in principle, to test Bell's inequalities and quantum mechanics at a larger distance. It is conceivable that the present theory is a local approximation, that works well over not-too-large distances.

FIG. 25. Optical harmonic generation. (See Schawlow page in color insert.)

Figure 25 shows an example of nonlinear optics. An intense beam of light from a laser comes in, a red beam. It goes through the crystal of potassium dihydrogen phosphate. The optical electric forces are so large that they're not negligible, even in comparison with the forces binding the atoms. The optical electric field changes the susceptibility of the material, and optical harmonics are generated.

Nonlinear optics can also be used to provide "squeezed light," because you can make a parametric amplifier which will amplify one phase of the light and not the other. It is interesting that the understanding of the quantum nature of light has developed very much in the years since lasers came along. To me, at first that seemed unnecessary, because with lasers, we have so many quanta, we don't have to worry about the graininess of the light. But on the other hand, it gives us tools where we can work with single atoms, and even with single quanta, and that has led to a strong theoretical understanding. Glauber, for instance, has done a lot of important work in that area, and developed a concept of a coherent state, which is like ordinary laser light (33). If you have a pure photon state, where you know the number of photons exactly, you can't know anything about the phase. With the coherent state, you have about equal knowledge of the amplitude and phase. Or, if you take two components in phase and 90 degrees out of phase, they have the same amount of quantum noise uncertainty. Even with an ideally noise-free laser beam, you have the irreducable limit caused by the vacuum fluctuations. The noise is the square root of the number of photons, and if you have few photons, then that's a bad noise.

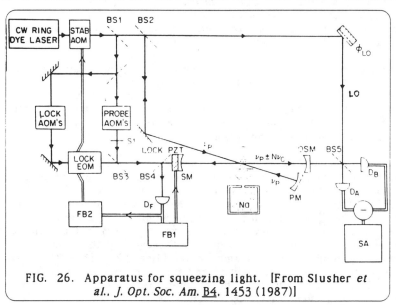

FIG. 26. Apparatus for squeezing light. [From Slusher *et al.*, *J. Opt. Soc. Am.* B4, 1453 (1987)]

But, what you can do now, is to take the noise from one phase and push it into the other. You can't just eliminate it entirely, but you can eliminate it from one phase by using a parametric amplifier, which is a phase sensitive amplifier. It amplifies for one phase and not for the other, because a beam of light from the laser produces a phase-dependent amplification by synchronously altering the optical susceptibility of some medium. It's too complicated to explain all these details from the paper of Slusher and his associates (34) but the schematic is shown in Fig. 26.

They were able to use a beam of sodium atoms pumped by a beam of light near resonance which changes susceptibility in synchronism with the light. Therefore they could amplify a probe with one phase, and deamplify the other phase; and so, deamplify the noise, even from the vacuum.

Now, this kind of squeezing may be important in the future for measuring gravitational waves, where you need ultimate sensitivity. Levenson and Wall and some others at IBM in San Jose have achieved squeezing with a glass fiber, using nonlinear optical effects (35). They have also done a quantum nondemolition measurement, where they could repeatedly measure the energy of a light beam, i.e. the number of photons, without destroying them. Their procedure introduced more phase noise, but kept the

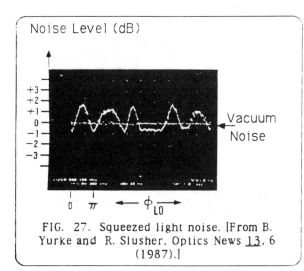

FIG. 27. Squeezed light noise. [From B. Yurke and R. Slusher, Optics News 13, 6 (1987).]

amplitude constant.

Figure 27 shows that Slusher and associates have actually reduced the noise below the vacuum level.

Finally, Fig. 28 shows an example of pure optics (36). It is an enhanced photograph of the recent supernova with a good telescope and a 256 by 256 photoncounting detector array. It doesn't usually pay to make a large diameter telescope for better resolution, for better resolution because the atmosphere scintillates so much and the image moves around. So, first they take a picture for only five milliseconds.

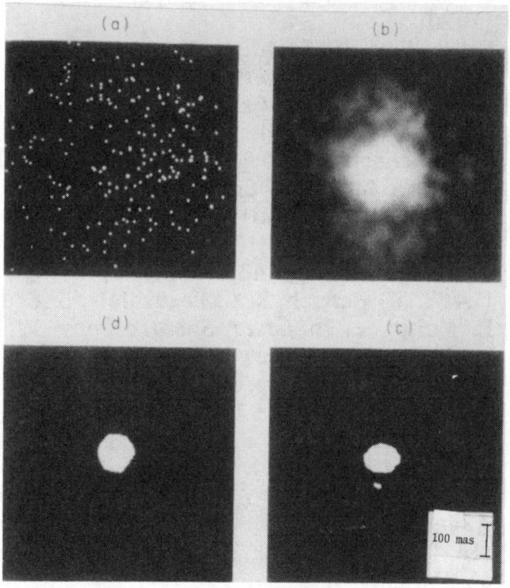

It's a pretty grainy picture, and you can't see much. They take 60 thousand of those, do a Fourier transform, as at the upper right, and superimpose the Fourier transforms, and then reconstruct it to get the image. And clearly there's a second object which is only one order of magnitude less bright than the supernova. It has been followed for several weeks at this time, and appears to be moving away at half the speed of light.

The resolution here is quite marvelous, a few arc-seconds. The best that one can do with ordinary astronomical photography is about a tenth of a second of arc. Thus we see that ingenuity can overcome a lot of limitations. Some limitations turn out to be more apparent than real, and perhaps even the ones that Professor Panofsky gave us on accelerators may be beaten somehow by clever people.

FIG. 28. Photo of SN 1987A with enhanced resolution. [From P. Nisenson et al., *Astrophys. J.* 320 L15 (1987).]

LESLIE L. FOLDY:

I'd like to thank Professor Schawlow for showing us how much there still remains to be done in optics, and I understand, is being done in optics, since Michelson. We've had a lot of delightful reminiscences from Dorothy Livingston this morning, a good deal of intellectual nutrition. I will now allow you to get some physical nutrition at lunch. I hope that the discussion there will be rewarding. Thank you.

REFERENCES

(1) D. C. Sinclair, *J. Opt. Soc. Am.* 66, 192 (1976).
(2) See, for example, *Ho Ho Hoffnung*, by Gerard Hoffnung, Harper and Row, New York, (1959).
(3) T. W. Hansch, M. D. Levenson, and A.L. Schawlow, *Phys. Rev. Lett.* 27, 707 (1971).
(4) C. Borde, *C. R. Acad. Sci, Paris*, 271, 371 (1970).
(5) T. W. Hansch, I. Shahin, and A. L. Schawlow, *Phys. Rev. Lett.* 27, 707 (1971).
(6) W. E. Lamb, Jr. and R. C. Retherford, *Phys. Rev.* 79, 549 (1950).
(7) L. S. Vasilenko, V. P. Chebotayev, and A. V. Shishaev, *J.E.T.P. Lett.* 12, 113 (1970).
(8) P. Zhao, W. Lichten, H. P. Layer, and J. C. Bergquist, *Phys. Rev.* A34, 5138 (1986); Phys. Rev. Lett. 58, 1293 (1987).
(9) F. Biraben, J.C. Gareau, and L. Julien, *Europhys. Lett.* 2, 925 (1986).
(10) R. G. Beausoleil, D. H. McIntyre, C. J. Foot, E. A. Hildum, B. Couillaud, and T. W. Hansch, *Phys. Rev.* A35, 4878 (1987).
(11) P. A. M. Dirac, *European Journal of Physics* 5, 65 (1984).
(12) S. Chu, A. P. Mills, Jr., and J. L. Hall, *Phys. Rev. Lett.* 52, 1689 (1984).
(13) S. Chu, A. P. Mills, Jr., A. G. Yodh, K. Nagamine, Y. Miyake, and T. Kuga, *Phys. Rev. Lett.* 55, 48 (1985).
(14) T. W. Hansch, R. G. Beausoleil, B. Couillaud, C. Foot, E. A. Hildum, and D. H. McIntyre, in *Laser Spectroscopy VIII*, W. Persson and S. Svanberg, eds., Springer Verlag (New York), 1987.
(15) M .E. Kaminsky, R. T. Hawkins, F. V. Kowalski, and A. L. Schawlow, *Phys. Rev. Lett.* 36, 671 (1976).
(16) N. W. Carlson, F.V. Kowalski, R. E. Teets, and A. L. Schawlow, *Opt. Commun.* 18, 1983 (1979).
(17) N. W. Carlson, A. J. Taylor, K. M. Jones, and A. L. Schawlow, *Phys. Rev.* A24, 822 (1981).
(18) W. M. Fairbank, Jr., T. W. Hansch, and A. L. Schawlow, *J. Opt. Soc. Am.* 65, 199 (1975).
(19) W. Neuhauser, M. Hohenstatt, P. E. Toschek, and H. G. Dehmelt, *Phys. Rev.* A22, 1137 (1980), and in *Spectral Line Shapes*, B. Wende, Ed., de Gruyter, New York (1981).
(20) T. W. Hansch and A.L. Schawlow, *Optics Comm.* 13, 68 (1975).
(21) S. Chu , L. Hollberg, J. Bjorkholm, A. Cable, and A. Ashkin, *Phys. Rev. Lett.* 55, 48 (1985).

(22) S. Chu, J. Bjorkholm, A. Ashkin, and A. Cable, *Phys. Rev. Lett.* 57, 314 (1986).

(23) E. L. Raab, M. Prentiss, A. Cable, S. Chu, and D. L. Pritchard, *Phys. Rev. Lett.* 59, 2631 (1987).

(24) A. Ashkin and J. M. Dziedzic, *Science* 235, 1517 (1987).

(25) G. Gabrielse, X. Fei , K. Helmerson, S. L. Rolston, R. Tjoelker, T. A. Trainor, H. Kalinowsky, J. Haas, W. Kells, *Phys. Rev. Lett.* 57, 2504 (1986).

(26) J. S. Bell, *Physics* 1, 195 (1965).

(27) S. J. Freedman and J. F. Clauser, *Phys. Rev. Lett.* 28, 938 (1972).

(28) A. Aspect, J. Dalibard, and G. Roger, *Phys. Rev. Lett.* 49, 91 (1982); *Phys. Rev. Lett.* 49, 1804 (1982).

(29) P. C. W. Davies and J. R. Brown, *The Ghost In The Atom*, Cambridge University Press, New York (1986).

(30) L. M. Ballentine, *Am. J. Phys.* 55, 785 (1987).

(31) John G. Cramer, *Rev. Mod. Phys.* 58, 647 (1986).

(32) C.O. Alley, O.G. Jakubowicz, and W.C. Wickes, *Proc. 2nd Int Symp. Foundations of Quantum Mechanics*, Tokyo (1986).

(33) R. J. Glauber, *Phys. Rev. Lett.* 10, 84 (1963).

(34) R. E. Slusher, L. W. Hollberg, J. C. Mertz, and J. F. Valley, *Phys. Rev. Lett.* 55, 2409 (1985).

(35) M. D. Levenson, R. M. Shelby, M. Reid, and D. F. Walls, *Phys. Rev. Lett.* 57, 2473 (1986).

(36) P. Nisenson, C. Papaliolios, M. Karovska, and R. Noyes, *Astrophys. J.* 320, L15 (1987). □□

WILLIAM FICKINGER:
Welcome back. It's my pleasure to introduce Robert Hofstadter, Professor Emeritus at Stanford University. Professor Hofstadter received the Nobel Prize in 1961 for his work in electron scattering, the measurement of hadronic form factors and nuclear radii. We're very pleased to have him with us today. He is going to chair this session for us. Professor Hofstadter.

ROBERT HOFSTADTER:
Thanks very much, Bill. I'm very pleased to be here today to celebrate this sort of hundredth anniversary of American Physics. I feel quite privileged to be able to do this, since I myself have been a physicist for a little over half that period, and I don't think I'll be around for the latter part of it. It's great to be here right now, it's great to see old friends and meet new ones, and see all the students who are going to become physicists in a little while, or maybe are already.

Anyway, I don't want to take any time away from the speaker, but I can't resist just making a remark or two about Hans Bethe, who is going to be our first speaker. When I was working on my PhD degree, which was in infrared physics, I never could get really very interested in the subject. I was always interested in nuclear physics. This was in 1937, or thereabouts, and I remember reading the articles in the *Reviews of Modern Physics* by Bethe, Bacher and Livingston. If you haven't seen them you should; they're still famous and wonderful to read. I feel that anything about nuclear physics that I know was initiated by those articles that I read. Professor Bethe is known for so many things that I couldn't possibly enumerate them here, but in particular he's known for working out the engine that drives the stars. I think he is going to speak about something related to that topic today. Hans Bethe of Cornell University.

The Life of the Stars

Hans A. Bethe

CORNELL UNIVERSITY

Ithaca, New York 14853

Well, I'm very happy to be here today and I was very happy about the introduction, except it made me feel just about a hundred years old. I was quite young when I wrote those articles.

We have enormous numbers of stars around us. Figure 1 shows you a typical galaxy; our own galaxy looks the same way. In the center there are lots and lots of stars, and then on the spiral arms there are somewhat fewer, but still very, very many. The total number is something like a hundred billion stars in one galaxy. There are however, also smaller assemblies of stars.

In Fig. 2 is shown is a picture of the

FIG. 1. S-24, the great galaxy in Andromeda; NGC 224, Messier 31. Copyright 1959 by Cal. Inst. of Tech. and Carnegie Inst.

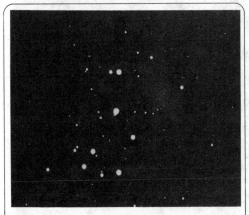

FIG. 2. The Pleiades.

Pleiades; this is called an open cluster. It is clear from the way they behave that they belong together. And the great advantage for the astronomer is that this shows that they are all about the same age.

In Fig. 3 is a more compact cluster, a so-called globular cluster. There are lots of those associated with the galaxy, mostly not in the plane of the galaxy but outside, above or below. They may have hundreds of thousands of stars. These globular clusters are particularly important because they are not only the same age, but they

FIG. 3. A globular cluster.

are also very old, and I'll come back to that.

So much about the phenomenology of the stars. Now, if one wants to say something about them, one better have an ordering principle. This was provided by the Danish Astronomer, Hertzsprung in 1911. Figure 4 shows the so called Hertzsprung - Russell diagram which plots in the vertical direction the luminosity of the star and in the horizontal direction their color. On the left side are blue stars, on the right side red stars. And contrary to the faucets in the hotel room, blue indicates hot and red indicates cool. Most of the stars are in one big strip, which you can see going from very luminous blue stars to faint red stars. This is called the main sequence, and by far the majority of the visible stars are in that main sequence. The sun is indicated near the middle, but somewhat down to the red end. Then you see on the right-hand side a lot of stars which are red and brilliant. These are known as the red giants. On the left-hand side, you see there are a few stars that are hot, but faint. These are the white dwarfs.

Figure 5 will show you the same thing, with some numbers attached to the horizontal. The surface temperature goes from about 3000 degrees to 25,000 degrees in this diagram. And of course, once you know the temperature and the luminosity, you can tell how big the surface of the star is because the light emitted per unit surface is just dependant on the temperature. The temperature you will derive from the spectrum of the star.

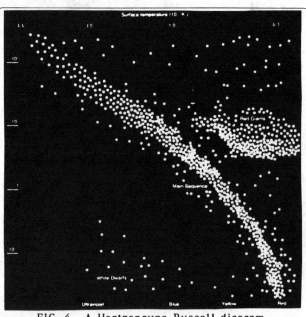

FIG. 4. A Hertzsprung-Russell diagram.

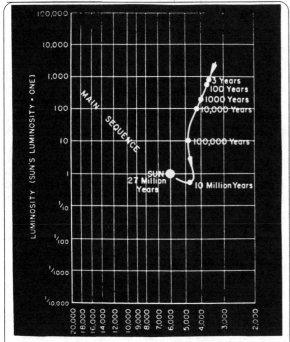

FIG. 5.   Hertzsprung-Russell diagram with scales.

Figure 6 gives you more information yet, namely the relation between the mass of the star, and the luminosity. Now, you will say it's easy to tell the luminosity, how about the mass; how can I tell that? Well, that is not so difficult because more than half the stars are in pairs, binary stars, which rotate around each other. You can measure the velocity of the two stars in their orbit, going either away from you or toward you, by the Doppler shift, and from the velocity as a function of time, and Kepler's laws, you can derive the masses of both the stars in the pair. Figure 6 shows you stars from a mass of about one tenth of the sun on the left to a hundred times that of the sun on the right. Then there is a curve labeled L, which gives luminosity, from 10^{-3} up to 10^5 of the sun's luminosity. You can see the luminosity increases much faster than the mass. The heavy stars expend their energy much faster than the light ones. Heavy stars commonly have a lifetime of only 10 to 100 million years, while the sun's lifetime is about 10 billion years.

Now we consider something more internal to the star. So far, I have told you what you can observe by looking at the star with a spectroscope. Eddington in the 1920's was able to unravel the interior of the stars. He did that by using very simple laws of physics: the simplest of them being hydrostatic equilibrium. This says that the pressure existing at any point in the star is equal to the weight of the column of mass which is above that point. It's the same as you have in the air causing a barometric change as you go to higher altitudes. The second point he needed to know is the equation of state: how does the pressure relate to the tem-

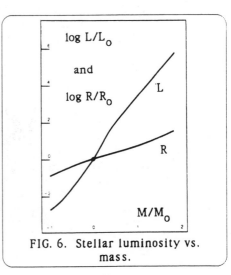

FIG. 6.   Stellar luminosity vs. mass.

perature and the density of the matter?

The third point is how energy is transported. Well, energy is transported by radiation that starts as very short wavelength radiation at the center, becoming visible light as you get out of the star. The energy is transported according to certain laws, and these laws were well known. It takes about 10 million years for radiation to get from the center of the sun where it is produced to the outside where we can see it. So what we see today is not what the sun is like today, but we see the energy it produced about 10 million years ago.

Now, Eddington got a very nice consistent picture, except for one thing: he made the assumption that the chemical composition of the sun was about the same as of the earth. And thereby, he found a very high temperature at the center of the sun, about 40 million degrees. Then other astronomers came along, particularly Stromgren, a Danish astronomer, who recognized that the sun and other stars, too, consist mostly of hydrogen. And that greatly lowers the central temperature to about 14 million degrees.

Eddington's theory gave us a picture of the inside of the stars, but he assumed that the energy is all produced near the center. He left open the question how that energy is produced. In the following table I outline the history of what people thought about stellar energy.

<u>HISTORY OF STELLAR ENERGY</u>

Infall of meteors	Mayer, Waterson, Thompson
Gravitation (1853-1905):	Waterson, Helmholtz, Kelvin, Newcomb
Radioactivity and dating (1905-19):	Soddy, Rutherford, Eddington
Element buildup (1920-24):	Perrin, Eddington, Russell
e-p annihilation:	Eddington, Jeans, Russell
Gamow theory of radioactivity Nuclear reactions (1929-):	Atkinson & Houtermans, von Weiszacker, Stromgren, Gamow & Teller, Bethe

The oldest idea (about 1850 or so) was that it may come from meteors falling into the sun. That of course is hopelessly insufficient. Very soon thereafter Helmholtz and Kelvin and two other people, had the more scientific idea that the various parts of the sun attracted each other, and that the self-gravitation of the star produces the energy. That worked all right, but it gave a time scale for the life of the sun of about 20 million years. And geologists knew very well that this couldn't be so. They knew how long it takes to deposit the various layers in the sea, and so the time must be longer than that.

Next came the discovery of radioactivity. Radioactivity permitted scientists to date old rocks. We know that the oldest rocks on the earth and on the moon are about four billion years old, and the sun is a little bit older than that. But that left the question: what is the source of the energy? Well radioactivity was there as a very natural idea for the source of energy. But it again doesn't give enough energy per unit time. So later on, in the early 1920's, Eddington and Russell (associated with the Hertzsprung-Russell diagram) had the idea that probably what was happening was a build-up of elements out of protons. That building up of elements then means that the energy is nuclear energy. It comes about by the collision of nuclei.

The trouble with that was that nuclei have tremendous positive charges, and positive charges repel each other. And the energy and velocity of the gas atoms were completely insufficient to overcome this electric barrier. That was solved by Gamow's theory of radioactivity. From 1929 on, everybody believes that the basis is the collision of nuclei with penetration of the potential barrier. The first paper on that was by Atkinson and Houtermans, and then others worked on this also. It was put into a very useful form by Gamow and Teller in 1937.

What had to be provided now is the specific nuclear reaction which makes the energy. The first idea, which was proposed by von Weiszacker, was that it is the reaction between two protons, hydrogen nuclei, making heavy hydrogen:

$$\underline{\text{Proton-Proton Reaction:}}$$

$$^1H + {}^1H \rightarrow {}^2H + e^+ + \nu$$

$$^2H + {}^1H \rightarrow {}^3He + \gamma$$

$$^3He + {}^3He \rightarrow {}^4He + 2\,({}^1H)$$

$$\text{Net Result:}$$

$$6\,({}^1H) \rightarrow {}^4He + 2\,({}^1H) + energy$$

This reaction has the difficulty that it can only go with the emission of a positron and a neutrino and that requires the weak interaction; that slows down the reaction very much. However, Critchfield and I, in 1938, calculated this reaction in detail and found that, with the assumed temperature at the center of the sun, it produces just the right amount of energy per unit time to explain the functioning of the sun. The result of this chain of reactions is that six hydrogen atoms successively are used; they make a ^4He nucleus releasing a great deal of energy, and two hydrogen atoms come back out.

This proton-proton reaction was very satisfactory for explaining the sun. However, as I showed you earlier, heavy stars produce energy at a much greater rate than the sun. The proton-proton reaction is not sufficient to explain that. One can calculate the central temperature of these heavy stars, and it is not much higher than that of the sun. So, we needed a reaction which increases very strongly, as the temperature increases. Such a chain of reactions is as follows:

Carbon-Nitrogen Cycle:

$$^{12}C + {}^1H \quad \rightarrow \quad {}^{13}N + \gamma$$

$$^{13}N \quad \rightarrow \quad {}^{13}C + e^+ + \nu$$

$$^{13}C + {}^1H \quad \rightarrow \quad {}^{14}N + \gamma$$

$$^{15}O \quad \rightarrow \quad {}^{15}N + e^+ + \nu$$

$$^{15}N + {}^1H \quad \rightarrow \quad {}^{12}C + {}^4He$$

This "Carbon-Nitrogen Cycle" consists of six successive reactions. The net result is again that four protons are combined to make a ^{4}He nucleus and lots of energy is released. The nice trick of this is that the ^{12}C which you start out with comes out of the chain of reactions; therefore we don't use up any carbon, we don't use any coal. We just use hydrogen which is the most abundant element in all the stars.

In Fig. 7 we show you how these two reactions depend on temperature. Temperature is given in millions of degrees. You see the flat curve gives the energy produced by the proton-proton reaction, which increases relatively slowly. The steeper curve is the carbon nitrogen cycle, whose energy production increases extremely fast. And that is satisfactory for the biggest stars alive. And so, this was a good explanation.

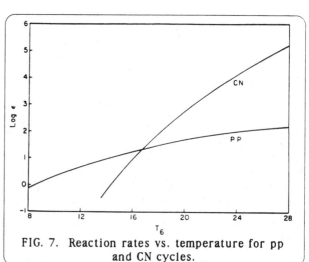

FIG. 7. Reaction rates vs. temperature for pp and CN cycles.

Now the question arises, how do the stars begin and how do they end? In Fig. 8 I show a picture taken in radio frequency of a certain region in the You see that there are a lot of radio waves coming from various regions, and then a particularly bright spot at the center. It is generally accepted that stars form where the mass is very dense. Therefore, they form generally in the spiral arms of the galaxy and probably also in

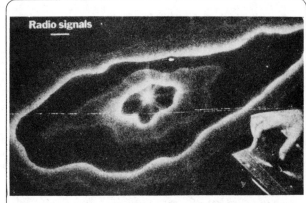

FIG. 8. Radio frequency photo. Illustration by Jim Ludke based on data of H. L. Taure from Cal. Inst. of Technology.

the center. (We don't see the center very well.) When there is a lot of mass around, then mass attracts mass and a central blob forms first. Then that attracts all the material around it. In the course of this, light is emitted. This light emission is shown on the Herzsprung-Russell diagram shown in Fig. 9.

You see that the newborn star starts at an extremely high luminosity, something like a thousand times that of the sun, and then as it contracts it emits less and less light. Finally it ends up very close to the main sequence. The time history shown is for stars of the same mass as the sun. This track of development in the diagram was derived first by Hayashi, the famous Japanese astrophysicist.

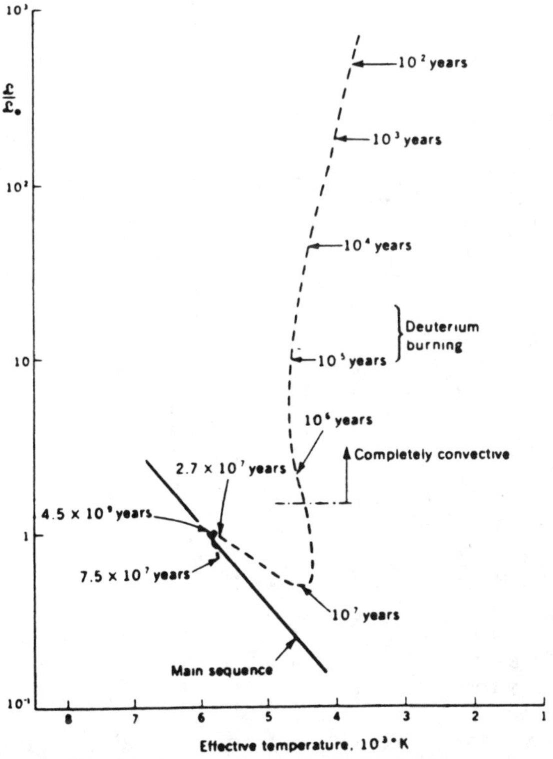

FIG. 9. Herzsprung-Russell diagram.

I now come to the problem,
what happens to a star at the end
of it's life? As I told you, the sun
will have a total life of about 10
billion years, 10^{10} years. It uses
up its hydrogen where it is hot,
namely near the center. It is
pretty well established that in the
sun and most other main-se-
quence stars, there is very little
mixing of the material in the interior of the star. So when you use
up the hydrogen in the center,
you terminate the energy production there. Once that happens,
there is no longer any pressure
developed in the gas near the center, and therefore gravitation has
nothing to oppose it. Gravitation
then will make the core of the star
contract very quickly to very
high density.

In Fig. 10 is shown the logarithm of the central density, relative to the density of water, of a
star which has used up its hydrogen. The horizontal axis is time.
It takes about 10 million years to
do this. You see that the central
density changes from about 40 to
something greater than 10,000
when this contraction happens.
And that is rather spectacular.
This figure and the following two
were calculated by Iben at the
University of Illinois.

When you compress a gas, you
always heat it at the same time.
Figure 11 shows how the temperature goes. Again the logarithm is
given. You start from about 25
million degrees, and the temperature goes up in very short order,
maybe one million years, to 100
million degrees. It increases
fourfold. Now, what good does that
do? The hydrogen is all used up.
But, there is still hydrogen outside
the core. And as the center contracts, also the region immediately

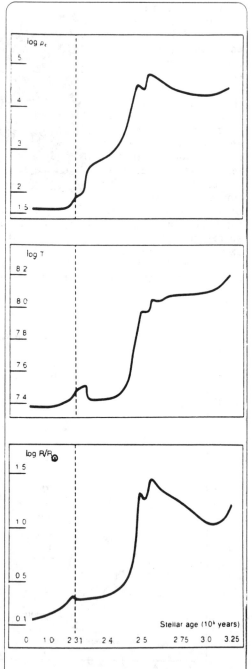

FIGS. 10, 11, and 12. Stellar central
density, temperature, and radius as
a function of time.

around the center will get higher density and higher temperature. Now you have high temperature in a shell around the core of the star. You now have a shell source of energy. And this shell source will provide just about as much luminosity as the core did before. Now what does the outside of a star do while all this happens in the deep interior? One would expect, if stars were reasonable things, that the outside would also contract, but not so at all.

In Fig. 12 is shown the radius of the star as a function of time, compared to the radius of the sun. You see the radius of the outside increases tremendously, from a little more than the radius of the sun (because we're talking about star three times the solar mass) to something like 20 times the radius of the sun: eight thousand times the volume. Well, this is very spectacular. Since the luminosity has remained more or less the same, the surface must get much cooler, because there is so much more surface. And so the star, once it has used up its hydrogen in the center, becomes a cool star, a red star, a giant star. So that's the way the giants come about. I showed you the giants on the Herzsprung-Russell diagram in Fig. 5.

Now the great question is, why does this happen? It doesn't seem reasonable. The point is, the core develops on it's own. It contracts; it has its own distribution of density, temperature, and energy production. But now a lot of energy is produced much closer to the center than it used to be, because that shell has contracted almost to the center. This energy finds it very difficult to get out of the star. The only way for the radiation to get out, is to have much lower density. So, around the energy producing shell, the star expands tremendously.

In Fig. 13 is shown the detailed development of a star from the main sequence to the giant. The stars increase in luminosity, and they become red, moving over to the right. Then they stay red and become more and more luminous. This is the characteristic development. This figure is from the book

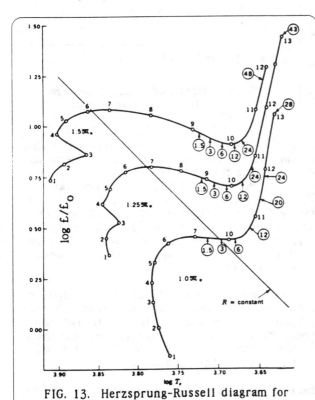

FIG. 13. Herzsprung-Russell diagram for sun-like star.

by D.D. Clayton [*Principles of Stellar Evolution and Nucleosynthesis*, University Chicago Press (1983)]. The numbers have something to do with the time at which it happens. The curves correspond to a star the mass of the sun, one and a quarter times the sun, one and a half times the sun, etc. But this is all done for a star which has the same composition as the sun. And the sun is very far from exhausting its hydrogen and therefore is not so interesting. What is interesting is what happens to old stars in a globular cluster.

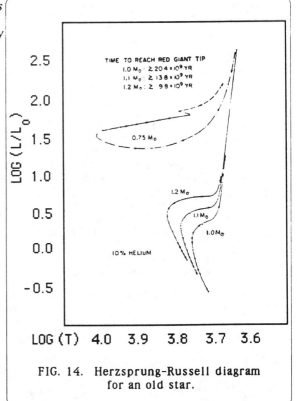

FIG. 14. Herzsprung-Russell diagram
for an old star.

In Fig. 14 is shown the same kind of development in a star which is typical in a globular cluster. Since these globular clusters are very old, they have a different chemistry from the sun. They have much more hydrogen and helium and much less heavier elements. But the general idea is still the same: you see the stars developing along the main sequence, moving over to the red, then becoming more and more luminous. That goes for stars the mass of the sun or greater. When the temperature in the center has become so high that helium starts to react, three helium nuclei collide together and make one carbon nucleus. That takes place at a temperature of about 100 million degrees at the center. Once that happens, the star suddenly goes back closer to the main sequence. It still has very high luminosity, but it moves back to the blue. Then the same thing repeats : the helium is used up, the helium becomes a shell source, and we go back to a red giant.

In Fig. 15 is shown the distribution on the Herzsprung-Russell diagram of stars in one of the globular clusters. There is indeed a horizontal branch leading over to the red, then going to very high luminosity (a hydrogen-shell source). Then they move back when the helium ignites, and back again to the red when the helium becomes a shell source. So, that is what happens to stars when they have used up their hydrogen.

Now, you may ask what happens later on? We have processed the hydrogen into helium, we have processed the helium into carbon, and so now the center of the star consists mainly of carbon and oxygen, and gradually, still heavier elements. The following table shows you the temperatures which are near the center of the star:

INTERNAL STELLAR TEMPERATURES (10^6 ° K)

Main sequence	10-30
Giants (M < 0.5 M_0)	30-100
Giants (upper branch)	150
End of evolution	< 5,000

The next table shows the temperatures needed to make certain reactions occur:

REACTION		TEMPERATURE (10^6 °K)
3 (^{4}He)	^{12}C	150
2 (^{12}C)	^{24}Mg	1200
2 (^{20}Ne)	^{40}Ca (also Si, S)	1900
2 (^{16}O)	^{32}S	3600
2 (^{28}Si)	^{56}Ni	5000

The reaction combining the alpha particles into ^{12}C requires a temperature of 150 million degrees. Two carbon nuclei can collide as the star gets still hotter near the center: that requires 1200 million degrees. And at two billion degrees, two neon atoms can get together to make calcium; at nearly four billion degrees, two oxygen atoms to make sulfur; and finally, two silicon atoms get together to make ^{56}Ni. Under normal conditions, ^{56}Ni is not stable, but it captures two electrons and makes the most abundant isotope of iron, which is ^{56}Fe. Now what happens after that?

Figure 16 shows the binding energy of nuclei as a function of the mass number from zero to 240. And you see binding energy increases from 1.1 MeV per nucleon in the deuteron up to a maximum of 8.7 MeV per nucleon

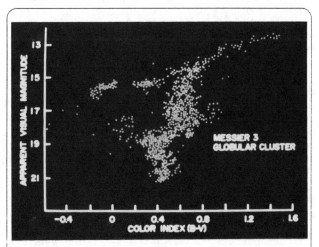

FIG. 15. Herzsprung-Russell scatter plot for a globular cluster.

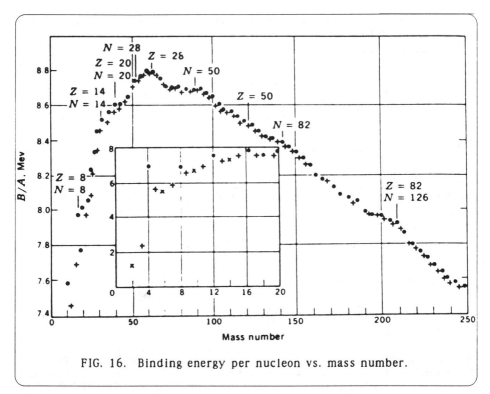

FIG. 16. Binding energy per nucleon vs. mass number.

at iron. You release energy as you get the hydrogens together to make helium, the helium nuclei together to make carbon, and so on up to iron.
When these nuclear reactions go on in the center of the star, you'll finally
make that center into iron. And once you have iron, you can't get any energy out of it anymore; iron is as strongly bound as any nucleus can be.

Therefore the same thing repeats: there is no energy production in the
center, gravitation dominates and makes the center collapse. This collapse
is slightly retarded, because there are electrons which make pressure even
at low temperature; you have a degenerate electron gas. But that's not important for our present problem. The core of such a star which has evolved
to having iron in its core can only collapse. And when it collapses, it does
so extremely fast. The core I'm talking about is about the mass of the sun.
It collapses in about one second to a radius of about 10 kilometers, to a neutron star. When that happens, the collapse has so much force that the outer
material then rebounds, and moves outward. Figure 17 shows the results of
a computation of how the material may move.

The different traces indicate the radial position of mass points originating from different radii. When the collapse has occurred place, the inner-
most ones stay at the center, but further out, they bounce back. This
amounts to a shock wave going outward, ejecting the material outside the
neutron star. In the center you have a neutron star, and on the outside,
you have material speeding out with great velocity. That is a supernova.

Nature was very kind to us, and gave us a supernova fairly close by last February, in the Large Magellanic Cloud at a distance of only 170 thousand light years. One of the important things about observing the supernova was that neutrinos were observed from it. Neutrinos, in contrast to light, can come out without hindrance, and so they give us the

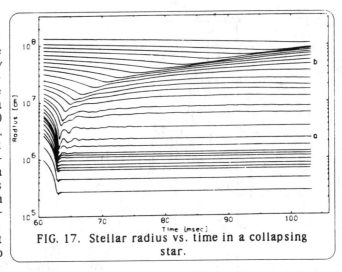

FIG. 17. Stellar radius vs. time in a collapsing star.

exact time when the supernova actually exploded. Then afterwards, you can see the light.

In Fig. 18 the dots indicate the observations of the light as a function of time. The lines are calculations. You can see that the agreement is not so bad. It has become somewhat better. Now, this is not the most modern calculation, but the agreement is clear enough. Toward the end, the light comes mostly from radioactivity, because what is formed in the star to be-

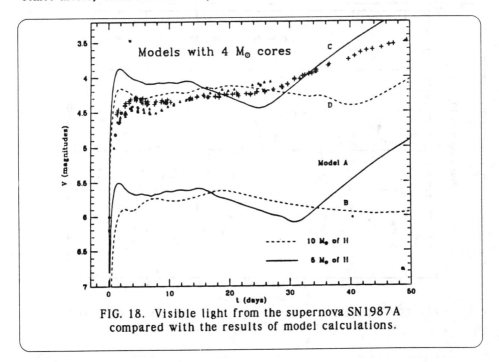

FIG. 18. Visible light from the supernova SN1987A compared with the results of model calculations.

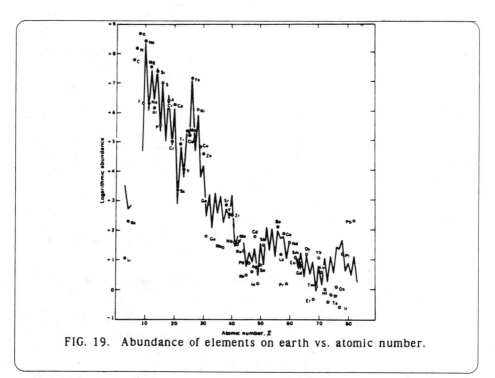

FIG. 19. Abundance of elements on earth vs. atomic number.

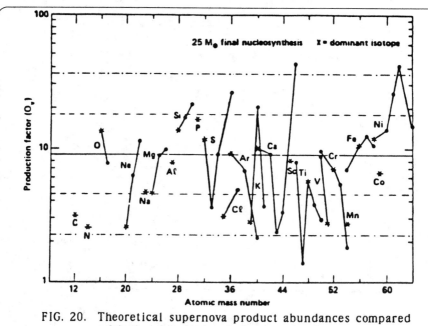

FIG. 20. Theoretical supernova product abundances compared
with the abundance of elements in the sun.

gin with is ^{56}Ni and the ^{56}Ni emits two positrons to make ^{56}Fe. That radioactive decay powers the supernova today. I can say that remarkable agreement has been achieved between the calculations and the observations, but not perfect agreement. We still have a lot to learn.

Well, this is the most spectacular part of the supernova, but there is one still more important thing. Namely, the supernova sends out all of the material, except for the one solar mass in the center, all the material which has been processed in the star, all sorts of interesting elements, from carbon on up. Figure 19 gives the distribution of elements in the earth as a function of their atomic number. In Fig. 20 is shown a comparison of the abundance of elements synthesized in supernovae as deduced theoretically from supernova theory with the abundance of elements in the sun. The big line in the center is where the elements should be. Even though the theory is not very accurate the differences are not great: Most of the elements and isotopes are within a factor 2 or 3 of the relative abundance which you see in the solar system.

Supernovae are terribly important, because they give us the material out of which the world is built, including ourselves. If there were no supernovae, we wouldn't be here either.

Thank you.

ROBERT HOFSTADTER:

The applause speaks for itself! That was a very beautiful talk on a beautiful subject. Thank you so much. □□

ROBERT HOFSTADTER:

Our next speaker is Fred Reines, who is going to talk about neutrinos from the atmosphere and beyond. And this will be a very fine connection with the talk just given by Hans Bethe. Again I don't want to take any time away from the speaker, but I can't resist making a personal remark or two.

Most of you know that Fred Reines and Clyde Cowan were the first ones to demonstrate the actual existence of a neutrino-like particle. I remember very vividly a conference at CERN in 1956, when Pauli received a telegram from Reines telling about the finding of the neutrino in that famous experiment. Pauli was obviously very happy about the whole thing, and I remember him waving the telegram in the air before this conference.

We're going to hear from Fred who in addition to his role in finding the neutrino has now connected it with events in a supernova as Hans Bethe has told you. It's also appropriate to say here that Fred is no stranger to this platform; in fact he used to sing on this stage as a member of the Cleveland Orchestra Chorus. You think you can give us a nice song, Fred? I won't ask him to do that really, I'll only ask him to give the next talk. And I think you will see how nicely it connects.

Neutrinos from the Atmosphere and Beyond

Frederick Reines

UNIVERSITY OF CALIFORNIA AT IRVINE

Irvine, California 92717

Thank you very much, Bob. I think what's been going on on this stage this afternoon is probably as exciting and artistic as anything heard here in many a year. It is a distinct and particular pleasure to join in this celebration at Case Western Reserve University of *Modern Physics in America*, because it affords an opportunity to describe work done at Case Institute which relates among other things to the beginnings of the new field of neutrino astronomy.

C. C. Giamati, W. R. Kropp, and F. Reines
Case Institute of Technology, Cleveland

Detector located: 2000 feet underground in the Morton Salt
Mine at Fairport Harbor, Ohio.

Scintillator Detector volume: 200 liters.
Cerenkov detector volume: 7 tonnes.
Primary motivation was to search for proton decay.
However, it had a sign on it declaring it was a:

SUPERNOVA EARLY-WARNING SYSTEM.

FIG. 1.

After the Pauli-Fermi neutrino was detected, it was clear that its extraordinarily weak interactions with matter made it an ideal particle with which to probe the interiors of stellar objects. You see in Fig. 1 a few of the facts in this history. Some Case students and I were involved in making a detector which we put two thousand feet under the ground in the Morton Salt Mines, now the Morton-Thiokol mine in Fairport Harbor.

The composite detector consisted of a 200 liter scintillation detector inside a Cerenkov detector which was all of seven tons. That doesn't seem much nowadays, as you will see we now have a Cerenkov detector which is some 7000 tons, but it was pretty big for those days. What did we build it for?

The primary motivation was to test the idea of baryon conservation (proton decay), a subject in which we had gotten interested several years earlier while at Los Alamos. I think that the time was the summer of 1954 when Maurice Goldhaber came by and said, "Say, you fellows have techniques

© 1988 American Institute of Physics

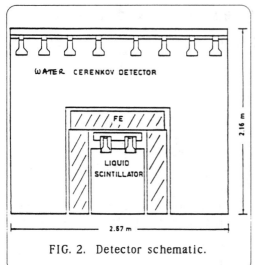

FIG. 2. Detector schematic.

which might be entirely appropriate to investigating these questions." Well, that intrigued some of us. Also, we were aware that the neutrino penetrated astronomical thicknesses of matter and was believed to play a vital role in the phenomena associated with supernova explosions. In fact we put a label on the tank, which read SNEWS, an acronym for SUPER NOVA EARLY WARNING SYSTEM! So we were optimistic, to put it mildly -- the detector was somewhat too small for that purpose. Figures 2, 3, and 4 show the detector.

The figures give you an idea of the scale. Kneeling by the detector is our engineer, Gus Hruschka, one of those gifted people who could make almost anything you could describe. The other figure happens to be my 6-foot tall son.

What physics results came from this early detector? As indicated in Fig. 5 we obtained lower limits on proton decay which were really quite stringent; up to as much as 4×10^{28} years depending on the decay mode assumed. We also used these data assuming neutrinos from the sun (as given by the boron-8 chain which had been described by Willie Fowler), to arrive at limits on the cross section. That was prior to some of the more current theories.

The interesting point about these neutrino reactions is that in order to understand, for instance the supernova, it is necessary to know these interaction cross sections. In the course of time we measured very many of these in the laboratory, including elastic scattering.

Well, that was the first of our forays into serious concerns regarding baryon non-conservation. We realized that we were limited by the detector size and set about considering how to increase its sensitivity. At the same time we became aware of the possibility that we might be able to detect neutrinos born in the atmosphere following the collision of energetic cosmic ray primaries

FIG. 3. Tank of main detector.

FIG. 4. Interior of tank.

with the atmosphere.

To pursue these goals, several of us got together and went to a mine in South Africa. It was the deepest mine in the world chosen to shield against the penetrating cosmic rays. We did a sequence of two experiments during the period 1964-1972. (Fig. 6.)

What were we looking for? Well, the cosmic rays strike the earth's atmosphere and make various secondaries -- kaons, pions, muons -- whose decays produce neutrinos. It was our goal to look for these "atmospheric neutrinos" and see what they were like; was theory correct, were there predictions to be checked?

Figures 7 and 8 give an idea of the experiment. You see at the upper right (in Fig. 7) neutrinos from the atmosphere. The neutrinos interact in rock, around the detector, and if their interactions give rise to muons, then the muons will be detected with a large area scintillator, located in the cavity of the rock.

The effective size of the detector was not simply the mass of those scintillation detectors, but the projected area times the range of the muon which penetrated the rock and then passed through the detector. And that of course made for a much greater effective mass, on the order of a thousand tons or so.

The first phase of the experiment was about 10,500 feet below the surface of the earth. This was the so-called 76th level experiment illustrated in Figs. 9, 10, and 11. The detector size

- lower limit on the proton lifetime

$$6 \times 10^{27} \text{ to } 4 \times 10^{28} \text{ years}$$
depending upon the assumed mode of decay

- upper limit on the solar neutrino flux (assuming V-A elastic scattering)

$$f(^8B) < 10^9 \; \nu_e/\text{cm}^2\text{sec}$$

to be compared with the then-predicted value of $(2.5\pm1) \times 10^7$ for ^8B neutrinos. (Current prediction is 4×10^6.)

- upper limit on the neutrino elastic scattering cross section (assuming standard solar model value for solar flux)

$$\nu_e e^- \rightarrow \nu_e e^-$$

$$\sigma_e \, (E_\nu > 8.5 \text{ MeV}) < 10^{-43} \text{ cm}^2$$

- In 1976 the companion process, elastic scattering of $\bar{\nu}_e$ on e^-, was measured at a Savannah River reactor with a detector designed at Case Institute.

FIG. 5. Physics results from Case salt mine detector.

M. Crouch, H. S. Gurr, T. L. Jenkins, W. R.
Kropp, F. Reines, G.R. Smith, H. W. Sobel
Case Institute of Technology, Cleveland
B. Meyer, J. P. F. Sellschop
*University of the Witwatersrand,
Johannesburg*

Located in the deepest mine shaft in the
world, near Johannesburg, South Africa.
Phase I - 1964 - 67
Phase II - 1967 - 72

FIG. 6.

was 4400 gallons of mineral oil
scintillator and at the time it was
the largest detector that had been
built. The rock temperatures were
very high and special ventilation
was required.

The detector was in a drift fifty
meters or so in length. This experiment was done, by the way, by
a collaboration of modest size;
you've seen essentially all the
names involved.

What did we learn from the

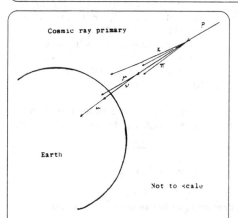

FIG. 7. Primary cosmic rays produce neutrinos in earth's atmosphere.

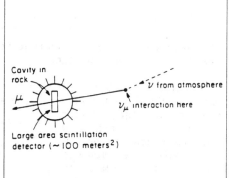

FIG. 8. Cosmic ray muon hitting
detector.

Detector location: 10,500 feet underground (at the 76th level).

Detector size: 4400 gals of mineral oil - the largest detector in the
world at that time.

Rock temperatures: 123 °F.

FIG. 9.   Phase I (level 76)
detector.

Phase I ERPM experiment? (Fig. 12)
Well, we actually saw these neutrinos,
and also obtained improved proton
lifetime limits, namely up to as much
as 8×10^{29} years. The lifetime results
depending again on the decay modes
assumed.   During this period we
heard that somebody had been seeing
some pulses and was attributing them
to neutrinos produced in a gravitational extraterrestrial collapse. We
looked at our data to see if there was
any connection between them. We
didn't see any, but that line of
thought may presage an interesting
development for the future.

Having done that, we said let's

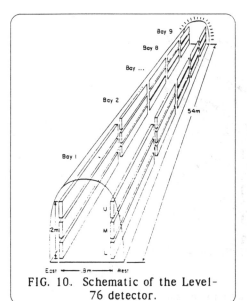

FIG. 10. Schematic of the Level-76 detector.

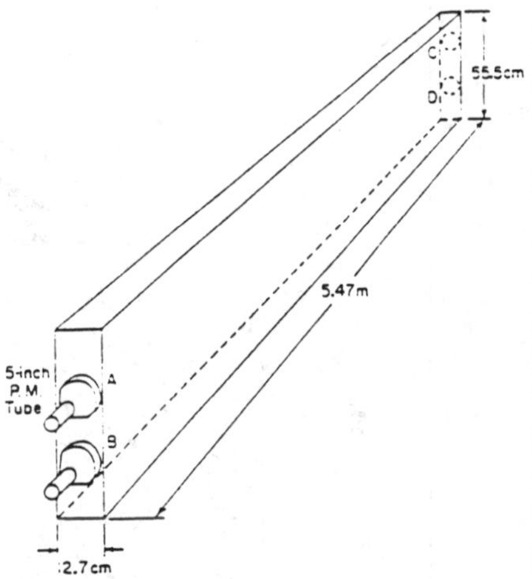

FIG. 11. Detail of detector element.

Detection of the neutrinos produced in our atmosphere by cosmic rays.

Nucleon-decay lifetime limits: $> 2 \times 10^{28}$ to 8×10^{29} years, depending on the decay mode assumed.
Upper limit on neutrinos associated with proposed gravitational pulses.

Fig. 12.

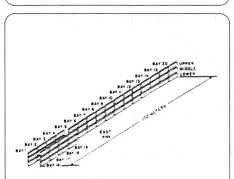

FIG. 13. Schematic of Level 77-detector.

learn a little more about the angular distributions of the atmospheric neutrino-produced muons. So, we went to another even deeper place (10,800 feet) in the same mine, dug yet another tunnel from scratch, and filled it with liquid scintillation slabs 174 square meters in area comprising some 20 tons of mineral oil based liquid scintillator, and viewed the tracks with a scintillation hodoscope having as its constituents some 48,384 flash tubes. The layout is shown in Figs. 13 and 14. This time we surrounded the trigger scintillators by flash tubes arranged in the hodoscope.

It was not a simple business, and it was complicated by the 1/2 hour travel time down to the mine. It was nice to get there though because it was air conditioned and not unpleasant to be in after the hot trip down. But one had the oppressive feeling that there was an awful lot of rock overhead.

Figure 15 shows the nature of the curves obtained. We have plotted the number of counts against the zenith

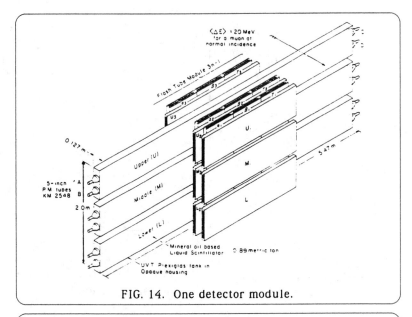

FIG. 14. One detector module.

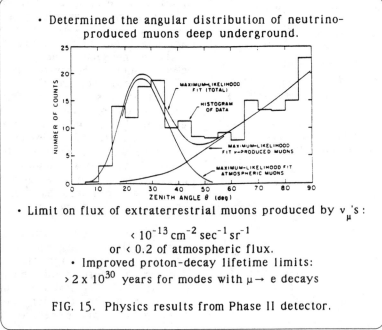

- Determined the angular distribution of neutrino-
produced muons deep underground.

- Limit on flux of extraterrestrial muons produced by ν_μ's:

$$< 10^{-13}\,\text{cm}^{-2}\,\text{sec}^{-1}\,\text{sr}^{-1}$$

or < 0.2 of atmospheric flux.
- Improved proton-decay lifetime limits:
$> 2 \times 10^{30}$ years for modes with $\mu \to$ e decays

FIG. 15. Physics results from Phase II detector.

angle. The angular distribution observed is that expected if the signal were
due to neutrinos.

Further, since we did the experiment at two different levels where the
cosmic rays were different, cosmic rays other than neutrinos were ruled

Irvine–Michigan–Brookhaven Collaboration (IMB)

W. Gajewski, K.S. Ganezer, T.J. Haines, W.R. Kropp, R. Miller,
L.R. Price, F. Reines, J. Schultz, H.W. Sobel, R. Svoboda
University of California, Irvine
J. Matthews, D. Sinclair, G. Thornton, J.C. van der Velde
University of Michigan, Ann Arbor
M. Goldhaber
Brookhaven National Laboratory
G. Blewitt
California Institute of Technology, Jet Propulsion Laboratory
C.B. Bratton
Cleveland State University
S.T. Dye, J.G. Learned
University of Hawaii, Honolulu
T.W Jones, M.S. Mudan
University College, London
D. Kielczewska
Warsaw University
M. Crouch
Case Western Reserve University
S. Errede
University of Illinois
H.S. Park
University of California, Berkeley
R.M. Bionta, C. Wuest
Lawrence Livermore National Laboratory
J.M. LoSecco
University of Notre Dame
D. Casper, A. Ciocio, R. Claus, S. Seidel, J.L. Stone, L.R. Sulak
Boston University
G.W. Foster
Fermi National Accelerator Laboratory
B. Cortez, E. Shumard
AT&T Bell Laboratories

FIG. 16. Salt-mine II experiment.

out as the source of the signals.

An independent experiment was subsequently done by the Kolar Gold Fields group, at a different depth. The agreement between the two groups supported the conclusion that we were both observing atmospheric neutrinos. I make this point because an independent experimental check is essential to a proof.

The next important advance was sparked by the emergence of the so-called minimal SU(5) theory which predicted proton decay rates accessible to experimental test. The enthusiastic support accorded this theory by the theoretical community persuaded the funding agencies to support experiments ex-

plicitly designed to test it. The group we formed to carry out the experiment called IMB (IRVINE /MICHIGAN /BROOKHAVEN) is listed in Fig. 16. Its size, thirty-odd people, was a large assemblage for many of us.

The detector designed and built by IMB, a ring-imaging Cerenkov, was unlike the earlier relatively small Case detector. It too was located in the Morton-Thiokol Mine near Cleveland. We made a search of many places around the U.S.A. and the world and we ended up coming home again, as it were. The detector and laboratory are in a specially built region, about 2,000 feet deep. The size is about 20 meters cubed, more or less; 7,000 tons of water, specially cleaned, and 2,048 eight-inch photomultiplier tubes located on its periphery.

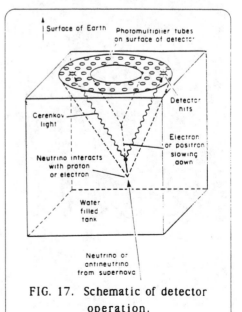

FIG. 17. Schematic of detector operation.

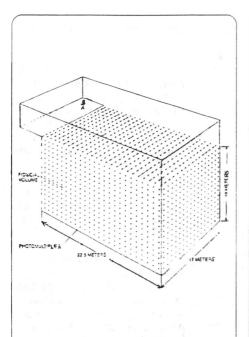

FIG.18. IMB detector schematic.

FIG. 19. The Digger.

FIG. 20. Water in IMB detector.

• Limit on p → e⁺ π⁰ (1983)

> 6.5 x 10³¹ yr

This important result ruled out the simplest grand unified gauge theory, SU5.

This result was improved to

> 1.2 x 10³² in 1985.

• Limits on 33 other proton and neutron decay modes, e.g.:

decay mode	lifetime (years)
p → e⁺ γ	>1.8 x 10³²
p → μ⁺ K⁰	>2.9 x 10³¹
n → ν K⁰	>1.0 x 10³¹
n → μ⁺ π⁻	>3.8 x 10³¹

• Limits on magnetic monopole catalysis of nucleon decay.
• Limits on n-n̄ oscillations
(ΔB = 2, where B = baryon number) in oxygen:

lifetime > 2.4 x 10³¹ years.

FIG. 21. Nucleon decay results.

Frederick Reines

Using atmospheric neutrinos:

Test of neutrino oscillations. (*Phys. Rev. Lett.* 54, 2299 (1985)) No evidence for oscillations observed. For maximal mixing, exclude the region:

$2.2 \times 10^{-5} < \Delta m^2 < 11.2 \times 10^{-5}$ eV^2

Limits on neutrino lifetime. (*Phys. Rev.* D35, 2073(1987)) $\tau p/m > 0.1$ seconds.

FIG. 22.

• Continuing study of possible relationships betgween deep underground muons and various astrophysical objects, e.g. Cygnus X-3 and Hercules X-1.

• Upper limit on steady flux of extraterrestrial neutrinos. [Astrophysical J. , **315**, 420 (1987)]:

$(0.5 \text{ to } 76.0) \times 10^{-4}$ cm^{-2}sec^{-1} depending on the source location.

• Observation of Supernova 1987A.

The observation of this supernova was the first definitive detection of extraterrestrial neutrinos and thus marked the birth of observational neutrino astronomy.

FIG. 23. Astrophysics results.

Figure 17 shows how a Cerenkov ring-imaging detector works. A relativistic charged particle produces a cone of Cerenkov light which is detected by photomultiplier tubes. By noting and reconstructing the hit-pattern pulse outputs and timing, the tracks of the particle can be determined.

Figure 18 is a sketch of the detector. It's very large as can be inferred from the figure. Well, you don't get a hole by just wishing, you've got to dig it. Figure 19 shows the machine with which we did so.

• Background for the 8 observed events is ≈2 per day or 1.3×10^{-4} per 5.6 sec.

• The probability that the 8 events are only accidental background is $(1.3 \times 10^{-4})^7 \approx 10^{-27}$.

• The neutrino pulse arrived at 23 Feb 1987 UT 7h 35m 41.37s, about 2.5 hours before the visual sighting of SN1987A.

• A search of the IMB data from May 1986 to May 1987 reveals no other such pulse sequences.

In view of the above, *these data are attributed to neutrinos from SN 1987A.* [*Phys. Rev. Lett.* 58, 1494 (6 April 1987)].

FIG. 24.

Figure 20 is a view of the detector filled with clear water. Visible are photomultiplier tubes and wave-length shifter plates designed to enhance light collection.

The figures show results on nucleon decay (Fig. 21) and neutrino physics (Fig. 22) obtained with the IMB detector. I should mention that these results ruled out minimal SU(5). This meant there was no longer any quantitative theory to deal with the problem of baryon decay. Also shown are some modes and associated limits. These negative results constrain theoretical models, a useful exercise. In this company we are particularly aware that a negative experiment can on occasion have a great positive effect!

In addition one can use this device to search for a relationship between underground muons and various astrophysical objects. There is as yet no definitive experiment in which positive results have been obtained, but one can hope to learn a little more about such connections -- in the form of limits on the steady flux of extraterrestrial neutrinos (Fig. 23). But by all odds the most interesting result to date is the observation supernova 1987A.

Immediately following Shelton's announcement of the supernova sighting, we and the Japanese in Kamioka began a detailed search of the data. We of IMB found a burst of 8 events in a 5.6 second period, a few hours (as Professor Bethe indicated it might be) prior to the visible supernova. Figure 24 presents some aspects of our data analysis. These events were "contained," i.e., were totally inside of the detector. This meant that they were

Standard supernova theory says:
• when a star goes supernova it should emit a very large number of neutrinos during the first few seconds;

• the energy released by SN1987A should have been about 10^{53} ergs, and more than 99% in the form of neutrinos, i.e. neutrinos carry off more than 1000 times the energy emitted in visible light!

FIG. 25

neutrinos. Kamioka found 11 events in 13 seconds. These two sets of results are believed to be consistent considering the characteristics of the two detectors.

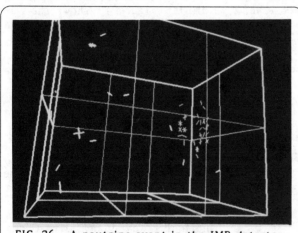

FIG. 26.   A neutrino event in the IMB detector.

There are various ways of describing the supernova data as seen by each detector separately. For instance you can ask "Was it accidental?" Then, "Is it systematic?" First, was it accidental? Given a steady rate of one or two per day, the probability of getting eight events in 5.6 seconds is really a reciprocal googleplex or so! There's no reasonable doubt, not even an unreasonable doubt, that this might be accidental.

But how to rule out systematic errors? If only one detector had seen it, I think we'd all agree that there'd be two opinions. One would say, "We have seen it, we know our own equipment, we've seen it." The other would say, "It isn't so, it's an artifact, it's systematic." Fortunately, there were two detectors, located on opposite sides of the world and they both saw a similar sequence of pulses within a period of a minute. And that, without any margin of reasonable doubt, settles it. (Papers describing both detectors are given in the *Physical Review Letters* of April 6, 1987).

Figure 25 summarizes some results from theory relating to the sightings. Most impressive is the fact that one can infer that most of the energy (>99%) involved in the supernova explosion comes out in the form of neutrinos. A corollary statement is that despite their weak interaction there are so many neutrinos emitted that a supernova located as far away as the orbit of Jupiter would be lethal to us even via neutrinos!

Figure 26 is a "picture" of a neutrino event obtained in the IMB experiment. Although the other seven events are persuasive to a seasoned observer who studies these kinds of reconstructions, this one is clear even to the uninitiated. And it fits the ring pattern beautifully.

You might ask what neutrino properties can be deduced from this sighting. One of them is based on the known distance to the supernova: the neutrino must live at least 170,000 years in order to reach us. In addition, one can make statements about the neutrino rest mass, and although these numbers are not by any means firm, one can be reasonably confident that it is less than 40, maybe less than 30 electron volts. And that's kind of interesting. These supernova results are perhaps beginning to be competitive with terrestrial experiments which are searching for evidence of neutrino mass. So we see that there are various neutrino characteristics that one can imagine pursuing.

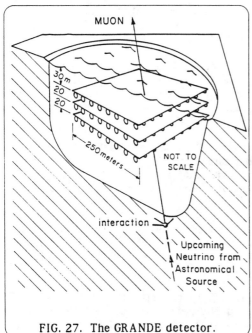

FIG. 27. The GRANDE detector.

What's next to be done now that a supernova has been seen by neutrino light? First off, what is the expected rate? Given a copious neutrino signal we could deduce both important features of the supernova and neutrino characteristics. We obviously need another supernova, and we have to be around and ready for it when it happens.

What is the rate of supernova occurrences in our galaxy as estimated from optical supernova counts in other galaxies? The rate of optically-visible supernovas could be one in 10 years if you're really quite optimistic. Actually one can ask, is it necessary that the light be visible for the super-

nova to occur? Or is the
light hidden because it is
somewhere in the center of
the galaxy, and you just
can't see that far. There
are arguments and debates
about this. It would be just
fascinating if the rate were
high enough so that some
of those here in the audience would be manning
one of these supernova detectors, in a giant world-
wide perpetual supernova
watch. Incidentally, it
would be fascinating to
search for coincidences between neutrino bursts and
gravity waves. It certainly
is possible at least for the
next few years for the current crowd to keep going,

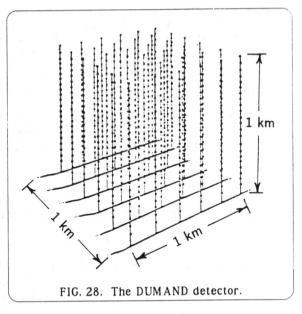

FIG. 28. The DUMAND detector.

and hopefully by then things will be different.

Now what other things can one think of? I want to close by mentioning
that there is a design and a proposal called GRANDE for a so-called GAMMA
RAY AND NEUTRINO DETECTOR. It would appear that one has a possibility of seeing very energetic neutrinos from outside our galaxy by means of such a
large detector as shown in Fig. 27.

It is a water Cerenkov detector about 250 meters on edge; is located at a
shallow depth; has several thousand photomultiplier tubes viewing parallel
layers of water from one level to another, and has directional resolution of
perhaps $1.5°$. A study indicates that one can make such a device, and it's
relatively inexpensive.

Another approach is a deep-underwater Cerenkov detector viewed by
photomultiplier tubes distributed throughout its volume. This approach,
known as DUMAND, is sketched in Fig. 28. So it promises to be a rich future
and we look forward to it with great interest.

I have attempted in this brief sketch to describe the rich variety of
physics results that has characterized the search for and study of atmospheric and extraterrestrial neutrinos with special emphasis on work at
Case Institute and some subsequent developments. In a mere 30 years the
neutrino has emerged from poltergeist to a tool with which to probe the
weak interaction and even more recently as a unique and powerful probe
of the interiors of the stars.

The promise for the future bespeaks well for the developments to be described in the meeting of the second hundred years of *Modern Physics in
America*.

ROBERT HOFSTADTER:

Thank you very much. We wish you well, and we wish others well who are looking for neutrinos in the same way. □□

KENNETH L. KOWALSKI:

It gives me great pleasure to present our next Chair who is an impressive condensed matter physicist. He's done me an enormous favor, namely I was afraid that I would have to introduce him when he was still at Cornell where with his multiple titles, I would surely get it wrong. So he's conveniently avoided that by taking a new job at the University of Maryland, where he is Professor of Physics [Physics Today 40, 97 (1987)], Michael Fisher.

MICHAEL E. FISHER:

Thank you, very much. I must start by correcting the Chairman. I'm actually a Professor of Physical Science and Technology at the University of Maryland. So, to those of you that know me, it means that I'm even more spread around. It is a great pleasure to be here and to see this. The older members of the audience don't need reminding that there used to be two institutions around here and some years back they were joined together in a time of difficulty and strain. It is a great pleasure for me personally to see how things are thriving and to be back to take part.

I would like to introduce the first speaker of this afternoon's session and I think the appropriate way of introducing is by answering the question that is surely in your minds. Peter Michelson *is* a relative of Albert Michelson. So, the answer is *yes*. Albert Michelson was Peter's great, great uncle. Peter was born out in San Francisco, took his BS at Santa Clara University in 1974, and went on to Stanford where he did his Doctorate in low temperature physics.

He's been on the faculty there since 1984 and has been interested in gravitational waves and that's in fact what he going to talk to us about. I was interested to discover he has an active collaboration with workers in China. He was out there for three and a half months and instead of the standard scientific tourism that many people do, he actually went back and worked with his collaborators. He's going to tell us this afternoon about the search for gravitational waves, something that many of us know has been going on for a long time; but we're getting closer and closer. He's going to "probe the dynamics of space time". Peter, its a great pleasure to introduce you.

The Search for Gravitational Waves:
Probing the Dynamics of Space-Time

Peter F. Michelson

STANFORD UNIVERSITY

Stanford, California 94305

It is certainly fitting at a symposium in celebration of an experiment that provided the base for Einstein's special theory of relativity to discuss with you the ongoing experimental effort to directly detect the gravitational radiation predicted by Einstein's general theory of relativity.

In fact, I'll talk about two experimental techniques. One of them is the resonant-mass detector effort, which originates with the work of Joseph Weber at the University of Maryland. The other technique, long baseline optical interferometry, is somewhat newer and is quite promising. It is especially appropriate to talk about the latter since it owes much of its heritage to the Michelson-Morley experiment.

Before describing the experimental efforts and the sensitivity of current detectors I will discuss what gravity waves are, how they interact with a detector and, the sources of gravitational radiation that we expect to be able to see.

We generally categorize these sources into three types: periodic sources, burst sources, and stochastic random background sources. This kind of categorization is in part due to the different techniques used to detect each type of radiation. Also, as you'll see, these different kinds of radiation come from different kinds of sources.

So I'll do that first and then I'll say something about the status of gravitational radiation detectors today, that is, the so- called free-mass, or laser-interferometric detectors, and the plans for building very long baseline, five-kilometer baseline interferometers.

Before proceeding further I must apologize to any of my colleagues in this field who might be in the audience, because I'm going to talk about a lot of work which is not due to me. Indeed there are are contributions from a large number of groups around the world. Table I is a list of these groups. As I proceed through the talk, I'll try to acknowledge their various contributions.

TABLE I.

RESEARCH GROUPS DEVELOPING GRAVITATIONAL RADIATION DETECTORS

RESONANT-MASS DETECTORS	INTERFEROMETRIC DETECTORS
Institute of Physics Beijing, China	Caltech-MIT
Louisiana State University	Max Planck Institute, Garching
Moscow State University	University of Glasgow
Stanford University	University of Paris, Orsay
University of Maryland	
University of Tokyo	Institute of Space and Aeronautical Science, Tokyo
University of Western Australia	
Zhongshan University Guangzhou, China	

Let me begin by describing a source of gravitational radiation which is very important. To date it probably provides the most convincing, but indirect, evidence for the existence of gravitational radiation, and indeed the gravitational radiation predicted by Einstein's general theory of relativity. The source I am referring to is the binary pulsar PSR 1913+16. This is a system presumed to consist of two collapsed objects, one of them being a radio pulsar (rotating, magnetized neutron star), gravitationally bound to each other in a binary orbit. The radio pulsar is a very good clock. By making timing measurements on that clock one can measure the orbital parameters of this binary system and it is found that the orbit is decaying with time. That is, the orbital phase shift is increasing. In Fig. 1 we see data from Joseph Taylor and his colleagues that clearly demonstrates this effect. The

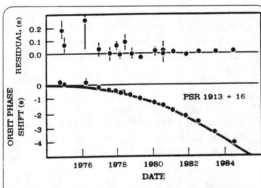

FIG. 1. Orbital phase shift vs. time for PSR 1913+16. The solid line is the phase shift expected from orbital decay due to gravitational radiation emission.
Adapted from J.H. Taylor and J.M. Weisberg, Astrophys. J. 253, 908 (1982).

solid line is the best fit prediction
(assuming general relativity) to
the orbital phase shift due to
gravitational radiation reaction.
The dots are the measurements.
This data represents a ten-year
observational effort and I think it
is really a remarkable achievement. There is little doubt in my
mind that this is clear observational evidence for orbital energy
loss due to the emission of gravitational radiation.

What are gravitational waves?
For this part of the discussion I
will take the naive point of view
of an experimentalist. A gravitational wave, if we're considering

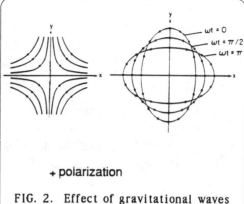

+ polarization

FIG. 2. Effect of gravitational waves
on a circular hoop of masses. The
tidal lines of force due to the passing wave are shown on the left.

detectors that are short compared to the wavelength of radiation, is really
nothing more than a time-dependent tidal force. For example, consider a
collection of masses as in Fig. 2 in the form of a necklace. If a gravitational
wave is incident on that system, one will observe (if the amplitude is large
enough) a relative motion of those particles as you see it on the right of the
figure. On the left is a plot of what the tidal lines of force look like due to
the wave. I should mention that these waves are transversely polarized. In
fact, there are two polarizations. The waves shown in Fig. 2 have the so-
called + polarization. To get the other polarization you just rotate the picture by 45 degrees.

Well, let me now describe in a little more detail how these waves interact
with various types of detectors. Let me begin with a free- mass laser interferometric detector pictured in Fig. 3. The detector consists of masses with
mirrors on them and a central beam splitter. (For example, this could be a
Michelson interferometer and one of Art Schawlow's lasers.) To avoid absolute length measurements the test masses (mirrors) are suspended to give
two perpendicular baselines. Imagine a gravitational wave incident normal
to the plane of the detector. What happens if we put a pulse of light into
this interferometer? The light beam will divide, part of it going down one
path, and part going down the other. What does the gravity wave do? If you
recall Fig. 2, what will happen is that during the first half cycle of the
gravity wave one arm of the interferometer, say along the y-axis, will get a
little bit longer, and the other one will get a little bit shorter. During the
next half cycle the opposite occurs. Thus the gravity wave induces a differ-
ential displacement in the lengths of the interferometer arms.

If we inject light pulses and measure the transit time down each arm of
the interferometer, we will find that there is a difference in the transit
times t that is proportional to the so-called metric strain h(t), a dimensionless quantity, which is of order $t/t_{transit}$, where $t_{transit} = 2L/c$. Now this of
course assumes that the interferometer is short compared to a wavelength
of radiation. One of the things you can do to increase the signal, since the

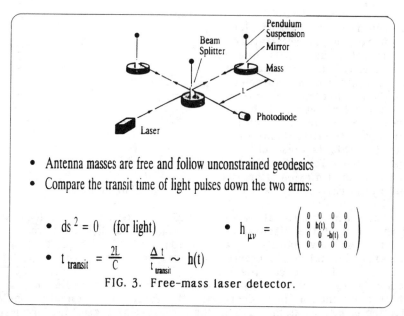

- Antenna masses are free and follow unconstrained geodesics
- Compare the transit time of light pulses down the two arms:

- $ds^2 = 0$ (for light)   • $h_{\mu\nu} = \begin{pmatrix} 0 & 0 & 0 & 0 \\ 0 & h(t) & 0 & 0 \\ 0 & 0 & -h(t) & 0 \\ 0 & 0 & 0 & 0 \end{pmatrix}$

- $t_{transit} = \dfrac{2L}{C}$ $\dfrac{\Delta t}{t_{transit}} \sim h(t)$

FIG. 3. Free-mass laser detector.

time delay is the measured quantity, is increase the overall transit time by making the interferometer longer. This works until you make the arms a quarter wavelength long; then you start getting cancellation of the signal. It doesn't do you any good to make it longer.

We expect to see interesting astrophysical signals at frequencies of the order of a kilohertz. A quarter wavelength at that frequency is about 150 kilometers; it is somewhat impractical to make an interferometer that long. We'll see there's a solution to this. I won't say what it is at this point, but nonetheless there are techniques for increasing the effective baseline.

There are other methods of making detectors; they don't necessarily have to be free-mass detectors. Figure 4 is a sketch of the kind of detector that Joseph Weber first pioneered, which, in its simplest form, consists of two point masses connected by a spring. Consider a gravitational wave incident on this system. Again the waves are transverse so a wave incident vertically in the figure can excite oscillations horizontally. In this configuration the mass quadrupole moment of the system interacts with the gravitational wave. The passing wave will deposit energy because it has to do work against these mechanical restoring forces of the spring. With suitable motion detectors one can detect the very small motion induced in this me-

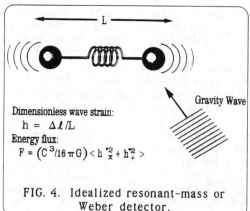

Dimensionless wave strain:
   $h = \Delta l / L$
Energy flux:
   $F = \left(C^3/16\pi G \right) < h_x^{\cdot 2} + h_+^{\cdot 2} >$

FIG. 4. Idealized resonant-mass or Weber detector.

chanical oscillator by the passing
gravitational wave. The size of
the motion is in the order of $\Delta L/L$
$= h$, where $h(t)$ is the dimensionless metric perturbation and $\Delta L(t)$
is the dynamic strain induced in
the antenna.

In anticipation of the discussion
of sources, I'll mention what the
typical values of h might be from
an astrophysical source. If we an-
nihilate one percent of a solar
mass in the galactic center and
convert it to gravitational radiation, by how much will the antenna length change? $\Delta L/L$ for that
kind of signal is the order of a few
times 10^{-18}. If the antenna is 3

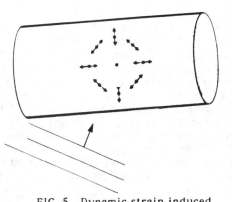

FIG. 5. Dynamic strain induced
in a extended-mass resonant detector by a gravitational wave.

meters long that is a displacement of only 10^{-17}m. So they're extremely
small displacements that we're trying to measure. In fact we do not expect
to see sources that are that efficient (except perhaps for coalescing binary
neutron stars). For example, the waves emitted in the core collapse of a supernova will probably be several orders of magnitude weaker.

This resonant-mass detector doesn't have to be made with point masses
and springs. In fact, what is usually done is to use an extended massive object which has several acoustic resonant modes. The massive object usually
takes the form of a solid right cylinder made from a material such as aluminum, with a fundamental longitudinal resonance frequency near 1 kHz. A
gravity wave interacting wave with such an antenna will induce a dynamic strain as indicated in Fig. 5. Typically we monitor the lowest fundamental longitudinal mode of the system in an effort to detect these waves. although, in principle, any mode that has a mass quadrupole moment can be
excited by a gravity wave pulse.

The size of a resonant-mass antenna is determined by the velocity of
sound V_s in the material used. Since V_s is always orders of magnitude less
than the speed of propagation of gravitational radiation, resonant-mass antennas are always much smaller than the wavelength of the radiation. I'll
come back this this point later when I discuss the sensitivity of resonant-
mass and free-mass detectors.

Before discussing the factors that determine the sensitivity of these detectors (i.e., noise), I will make some remarks about sources. Rather than
telling you all the mathematics of it, I think its easier just to do some simple
back of the envelope estimates. There are very sophisticated computer calculations done these days and I wish I had the time to tell you about some of
them.

If one makes a simple estimate, one finds that the luminosity of a typical
source is

$$L \sim (c^5/G) \, (R_g/R)^2 \, (v/c)^6 \quad \text{with} \quad R_g = G M/c^2$$

The luminosity scales as the quantity c^5/G which has dimensions of luminosity. This combination of fundamental constants is multiplied by two dimensionless numbers: i) the square of the gravitational radius R_g of the object of mass M, divided by the characteristic size of the source R and, ii) $(v/c)^6$ where v is a typical internal velocity in the source that is giving off the gravitational radiation. Immediately you can see that compact relativistic sources are the strongest and most luminous sources of gravitational radiation: things such as collapsing cores of supernovae, and coalescing binary systems, gravitational collapse, exploding galaxies, and so forth. Notice that the constant c^5/G appears to be the maximum intrinsic gravitational luminosity that can be radiated by a source with a mass M (and energy Mc^2). It is very easy to construct an elementary argument that demonstrates this.

TABLE II. BURST SOURCES OF GRAVITATIONAL RADIATION

Collapse and bounce of supernova cores:	$f \sim$ 10 - 10,000 Hz.
Corequakes in neutron stars:	$f \sim$ 100 - 10,000 Hz.
Births of black holes:	$f \sim$ 10,000 Hz(Mo/M)
Collisions between black holes and neutron stars:	$f \sim 10^{-4}$ - 1,000 Hz.
Final inspiral and coalescence of a compact binary system	$f \sim$ 100 - 3000 Hz.

Most of the effort in building gravity-wave detectors has been centered on trying to optimize them to detect broadband burst sources. I've listed in Table II a number of such sources and the characteristic frequencies of the radiation one expects. From a supernova for example, the frequencies of the radiation could be anywhere from about 10 to 10^4 Hz. 10^4 Hz is the frequency characteristic of the radiation emitted during the gravitational collapse of a one-solar-mass core. The efficiency of gravitational wave (GW) generation in a supernovae core is highly uncertain. From model calculations it is clear that the efficiency depends crucially on the angular momentum of the core. A spherical collapse produces no gravitational radiation, while the most favorable outcomes yield dimensionless metric perturbations h approaching 6×10^{-19} for a source located at the distance of the Large Magellanic Cloud (the location of SN 1987a). The radiation flux depends crucially on the degree of nonaxisymmetry in the collapse and also on the speed of collapse. For example, if the total net neutrino dissipation rate remains small, then GW efficiencies can rise as the core cools and significant nonaxisymmetries can grow as a result of successive bounces. However, if the supernova collapse is hot and collapse is slowed because of resistance provided by thermal pressure, GW emission will be reduced., The physics is very complicated and a clear picture will probably emerge when gravity waves are detected from a collapse.

Its obviously nice to detect gravity waves, but can we do anything more with them? What can we learn about these sources by measuring h(t)?

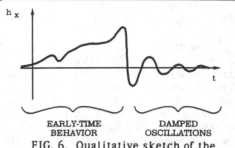

In Fig. 6 I have qualitatively sketched a typical wave form one might expect from the violent collapse of an object that would produce gravitational radiation. (This again is with apologies to theorists and numerical relativists who nowadays calculate these things in more detail.) For example, think of a scenario which produces a black hole in the final state. For example, consider a binary

FIG. 6. Qualitative sketch of the waveform of the gravitational radiation emitted when a black hole swallows a neutron star. Adapted from K.S. Thorne, *Rev. Mod. Phys.* 52, 285 (1980).

system with a neutron star orbiting a massive black hole. Eventually the orbit decays (just like the binary pulsar orbit) and the neutron star spirals into the black hole. The so-called early-time behavior gives us information about what's happening to the black hole while the neutron star is falling in; is the neutron star tidally disrupted; does it orbit around the black hole before it falls in, and so forth.

The actual coalescence involves some very complicated physics. However, late-time behavior is expected to be very simple. One expects to see a damped oscillatory motion that is a property only of the final black hole. These oscillations are the so called quasi- normal mode oscillations of the

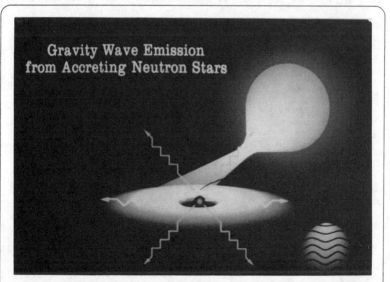

FIG. 7. Binary system with a neutron star accreting matter from companion star that has filled its Roche lobe.

hole. By measuring the frequency and the damping time we can, in principle, measure the mass and the angular momentum of the black hole.

There are other kinds of sources we could expect to see; so called continuous-wave sources. The binary pulsar PSR 1913+16 is an example. I've illustrated another source in Fig. 7 with a cartoon. In some binary systems containing neutron stars and companion stars that have overflowed their Roche lobe, mass transfer from the companion star to the neutron star is occurring. Because the matter has appreciable angular momentum, a flat accretion disk will form around the neutron star. Viscosity in the disk allows the matter to eventually lose sufficient angular momentum that it can accrete onto the neutron star's surface. These viscous forces also heat the disk to a sufficiently high temperature that the system emits X-rays. If the neutron star has a weak magnetic field the disk can extend down to the surface of the star and the accreting matter can spinup the neutron star to a very short period the order of a millisecond. In that regime it's possible, but not at all proven, that these stars will be subjected to instabilities due to viscous dissipative forces and gravitational radiation that will lead to a nonaxisymmetry of the star, which in turn leads to the emission of gravitational radiation. In this steady state, angular momentum added to the star by accretion is radiated away by gravitational wave emission. The resulting GW signal is expected to be somewhere in the frequency range 200 Hz to 800 Hz. The nonaxisymmetric distortion of the star associated with GW emission will also affect the accretion process producing a weak but perhaps detectable modulation of the X-ray flux at the same frequency. Thus we have a star that is simultaneously an X-ray and gravity-wave pulsar. Detection of weak X-ray pulsations at these relatively high frequencies will require a very large area X-ray detector with a very high time resolution. Such a detector, known as the X-ray Large Array (XLA) has been proposed, and is currently being studied, for deployment on the NASA space station in

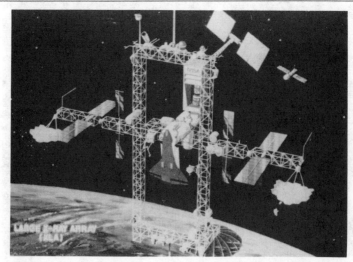

FIG. 8. Artist's conception of proposed large area X-ray detector array deployed on the NASA space station.

the 1990's. (See Fig. 8.)

The final category of sources that I will mention is stochastic radiation that is randomly distributed in space and time. Table III lists some of the possible sources. This radiation will be very difficult to directly detect and will require the operation of detectors with exquisite sensitivity at low frequencies in order to be observed. Stochastic GW radiation may prove to be the ultimate probe of the structure of spacetime in the early universe.

TABLE III. STOCHASTIC GRAVITATIONAL RADIATION SOURCES

Radiation from the big-bang singularity
First-order cosmological phase transitions
Formation and coalescence of black holes formed
from population III stars
Speculative objects such as cosmic strings

Well, now I'd better say something about the detectors themselves and what limits their sensitivity. The signal is only half the story; the experimentalist, of course, has to worry about a number of practical problems due to various noise sources including those that I have listed here:

Thermal noise
Seismic and mechanical noise
Cosmic ray showers
Gravity gradient noise
Heisenberg uncertainty principle

Thermal noise and seismic noise are among the most important noise sources to contend with at present. In principle they can be eliminated by a variety of experimental techniques. In the future (maybe not so distant) we will have to contend with more fundamental limitations that come from quantum mechanics. If a quantum-limited linear readout is used to monitor the relative positions of the detector masses in the case of an interferometric detector or to monitor the amplitude and phase of the fundamental longitudinal mode of a resonant-mass detector, then the limiting sensitivity can be estimated by the usual Heisenberg uncertainty principle arguments as

$$h \sim (1/L) \, (\hbar \tau / M)^{1/2}$$

where τ is the duration of the gravitational-wave pulse, and L and M are the length and mass of the antenna, respectively. This formula applies to both kinds of detectors. It would appear that by making L as large as we like, we can get any sensitivity we want. Unfortunately this is not so. The length of a resonant-mass antenna must be one half of an acoustic wavelength $V_s \tau/2$, while the maximum length of a free-mass antenna is one half of the wavelength of the gravitational wave $c \tau/2$. Thus the quantum limit of a free-mass interferometric detector is a factor of roughly $V_s/c \sim 10^{-5}$ smaller than the quantum limit for a resonant-mass detector. As we will

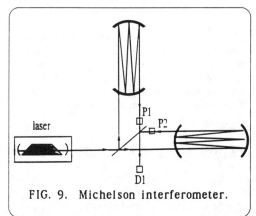

FIG. 9. Michelson interferometer.

see, the experimental requirements to actually reach these
quantum limits are quite demand-
ing and very different for each of
these two detectors.

I will discuss the noise limitations in free-mass interferometric
detectors first. Figure 9 shows the
kind of interferometer being developed in a number of laboratories around the world, including
the Max Planck Institute in Mu-
nich and MIT. It is known as a
Michelson delay-line interferometer. It consists of a marriage
of laser technology and high reflectivity mirror technology with the basic
idea that was developed by Albert Michelson more than 100 years ago. In
fact the technique of increasing the effective baseline to the optimum by
folding the beam and back forth between the mirrors is the same technique that Michelson used to improve the sensitivity of the 1887 experiment.

In Fig. 10 we see the expected contributions of various noise sources as
functions of frequency. These estimates are for a 5-kilometer interferometer using 100 watts of incident laser power. Note that the noise sources scale
in different ways with frequency. Consider first the limitation due to what
we call "Poisson" noise that arises
from the quantum nature of the
electromagnetic field and the unavoidable quantum fluctuations of
the number of photons detected
per unit time. This noise is sometimes referred to as photon counting noise. For a fixed laser power
this noise source increases with
frequency (of the gravity-wave
signal) since higher frequency
corresponds to shorter detector
integration times. The fractional
fluctuation in the number of photons detected during an integration time equal to $1/f$ is $(h/ P f_p)f$.
This noise can be reduced by increasing the laser power P. The
laser frequency is f_p. If we increase the laser power too much
then eventually we must contend
with the quantum noise from the
fluctuating number of photons
bouncing off the mirrors and im-
parting an uncertain momentum

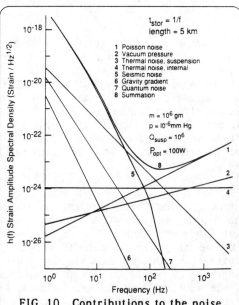

FIG. 10. Contributions to the noise
spectrum vs. frequency for a 5 km
laser interferometer.

to the mirror masses. This is shown in Figure 10 as quantum noise. Sometimes this noise source is referred to as "backreaction" quantum noise. This noise source is less at higher frequencies, scaling as $(h\ P\ f_p)1/2\ /\ (M\ L^2\ f^3)$. Notice that the quantum noise contribution increases with increasing laser power and is reduced by increasing the interferometer baseline L.

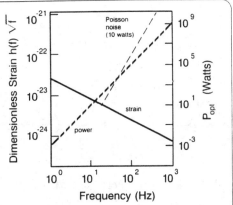

FIG. 11. The quantum limit of a free-mass laser interferometer detector (solid line) and the laser power required to reach this limit (heavy dashed line). The dashed line labeled Poisson noise is the strain noise level assuming that only 10W of laser power is injected into the interferometer.

In order to reach the quantum limit implied by the Heisenberg uncertainty principle, enough laser power must be injected into the interferometer so that the photon counting noise contribution equals the noise contribution from the backreaction quantum noise. Figure 11 shows the laser power required, as a function of gravity-wave frequency, for a 5 kilometer baseline. Also shown is the resulting dimensionless strain noise spectral density at the quantum limit. The required laser power scales as f to the 4th power. Thus it is very difficult to achieve the quantum limit at high frequencies. At low frequencies the laser power requirements are reasonable but thermal and seismic noise eventually intervene to limit sensitivity. By the way this is where photon squeeze states could help; not by allowing us to beat this quantum limit but by reducing the laser power required to reach it.

At the present time the laser powers actually being used in interferometers are the order of 20 milliwatts. This is much less than the power required to reach the quantum limit. Thus the most important noise contribution (at least in the vicinity of 1 kHz) comes from the Poisson photon counting noise. The noise performance achieved in prototype interferometers that are between 30 and 40 meters in length is approaching the Poisson noise limit for the laser powers being used. The rms dimensionless strain noise (in a bandwidth of about 1 kHz) is about $h \sim 10^{-17}$ in these detectors. Eventually, with a 4 km baseline and about 100 watts of laser power, the proposed LIGOs (Laser Interferometer Gravity Wave Observatory) could achieve $h \sim 3 \times 10^{-22}$ for detection of kilohertz bursts of gravity waves. Similar facilities are in the planning stages in Germany, France and Great Britain.

I should mention that some of these designs are based on Fabry- Perot cavity interferometers. These are being developed principally at Cal Tech and the University of Glasgow. The basic configuration of these detectors is shown in Fig. 12. Notice that the optical beams overlap each other within each optical cavity.

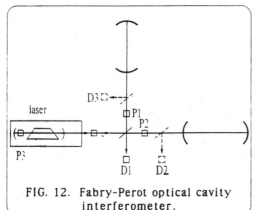

FIG. 12. Fabry-Perot optical cavity
interferometer.

For the remainder of my talk
let me summarize the current situation with resonant-mass detectors and mention some of the future plans. I certainly feel more
at home telling you about these
detectors because I have been
working on them for the past several years at Stanford University.
Figure 13 is a photograph of the
cryogenic resonant-mass detector
that we have developed and operated at Stanford. With apology to
the other groups listed in Table I,
I will mainly discuss the Stanford
detector. The current dimensionless strain noise level of this detector is h ~
10^{-18} for detection of kilohertz bursts of gravitational radiation. To understand the sensitivity limitations of these detectors we must again consider
the various noise sources that I listed earlier.

This antenna has been operated at 4.2 K with a dimensionless strain sensitivity of 10^{-18}. A second generation ultralow temperature detector, operating at 40 mK, will
have a sensitivity of
better than 10^{-20} at
the noise level.

It is sometimes use-
ful to characterize
the response of a
resonant-mass an-
tenna in terms of a
frequency- depen-
dent cross section
$\sigma(f)$. The energy
deposited in an an-
tenna originally at
rest is

E - E(f) σ(f) df

where E(f) is
the energy spectral
density of the signal
pulse and f_a is the
antenna mode fre-
quency. This ap-
proximate equality
holds if the pulse du-

FIG. 13. The 4800 kg resonant-mass detector located at Stanford University.

ration is much less than the ring-down time of the antenna. $\sigma(f)$ is a
sharply peaked resonant function centered at $f = f_a$ with a width at half
maximum of $(2f_a / Q)$, where Q is the mechanical quality factor of the antenna. This behavior of the cross section seems to imply that a high-Q antenna will have a very narrow bandwidth, but this is not the case. Both the
frequency spectrum of the strain signal induced in the antenna by a gravity wave pulse and the thermal Brownian motion noise of the antenna have
the same resonant response in the vicinity of the antenna's resonant frequency. Thus the signal-to-noise ratio (which is the important quantity) is
not bandwidth limited by the resonant behavior of the antenna thermal
noise. A sensitivity and bandwidth restriction comes from the electrome-
chanical transducer-amplifier readout. Indeed the thermal noise contribution scales as T/Q. Thus a high-Q, low-temperature antenna is very desirable. With a multimode resonant mechanical matching network, band-
widths of the order of $\Delta f = 0.25 f$ should be possible. In the present antenna at Stanford, which uses a matching network with only a single resonant
element, a bandwidth of 13 Hz has been demonstrated. The antenna mechanical Q is 5×10^6.

The noise contributed by the transducer-amplifier readout comes from
two essentially independent sources that are usually referred to as broadband amplifier noise (that simply adds to the detector output) and back reaction amplifier noise. These noise sources are unavoidable. In a properly
impedance-matched detector, these two noise sources make roughly equal
contributions to the overall noise budget. Even with a perfect amplifier
there is quantum noise from these sources that enforces the Heisenberg
uncertainty principle limit.

In order to detect the vibrations from these detectors, one
needs to have an electrome-
chanical transducer that converts the mechanical oscillations of the antenna into an
electrical signal that can be
amplified and then detected.
In our detector at Stanford, the
transducer consists of a me-
chanically-resonant superconducting diaphragm coupled
to the antenna. This transducer is illustrated schematically
in Fig. 14. Motion of the diaphragm modulates the induc-
tance of a superconducting
circuit in which a persistent
current has been stored. Because of magnetic flux conservation, the modulation induces
a current proportional to the
mechanical displacement to
flow in a coil that is coupled to

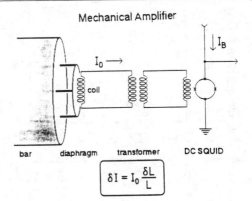

FIG. 14. Superconducting electrome-
chanical motion transducer. Motion of
the bar antenna is coupled to the me-
chanically resonant diaphragm. The
diaphragm is a superconducting
groundplane. Its motion modulates a
supercurrent that is stored in the circuit as shown. The modulation is coupled to a DC SQUID amplifier and then
detected.

the input of very sensitive SQUID (Superconducting QUantum Interference Device) amplifier. These devices, based on the Josephson effect, are the lowest noise amplifier's available in this frequency range (ie. around a kilohertz). Indeed nearly quantum limited performance of SQUIDs has been demonstrated.

With this kind of technology, we've achieved a strain sensitivity at the noise level of the current version of the detector of 10^{-18}. Recently, this detector was operated in coincidence with similar detectors located at Louisiana State University and at CERN in Geneva. (The CERN detector is operated by the University of Rome.) It is necessary to operate detectors in coincidence in order to discriminate against local disturbances that could excite the detector. A signal due to gravitational radiation would be registered almost simultaneously in more than one detector. While no statistically significant coincidences were observed, the observation carried out with these detectors has provided the lowest upper limit on the flux of kilohertz bursts of gravitational radiation that may be incident on the earth. This upper limit is shown in Fig. 15. During the this observation the noise in both the Stanford detector and the University of Rome detector departed from ideal Gaussian behavior. Indeed during an earlier observing period in 1982, the noise output from the Stanford detector was almost ideally Gaussian. The curve labeled Stanford(1982) shows the upper limit obtained from the 1982 data. The dotted line shows the upper limit that could be obtained with two such detectors oriented in the same direction and limited by Gaussian noise only. With a pair of second generation detectors that are now being constructed to operate at 40 mK, the sensitivity of two-fold coincidence observations will be improved by at least 2 orders of magnitude in strain (or 4 in energy). The lower abscissa demonstrates what such an improvement would mean in terms of detecting a supernova burst in the LMC (the location of SN 1987A).

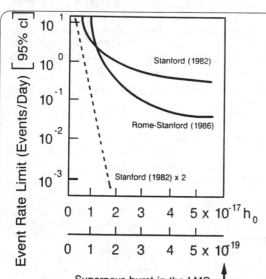

FIG. 15. Logarithmic plot of the upper limit on the rate and strength of gravitational wave pulses reaching the earth during observations with cryogenic resonant-mass detectors located at Stanford University and at CERN. The curve labeled Stanford (1982) was a previous upper limit obtained with the Stanford detector alone. See text for more details.

Before I run out of time, let me summarize the direction that I think developments with resonant-mass detectors will take in, say, the next 5 years. First of all, within the next two years we will have second generation antennas operating at about 40 mK (using dilution refrigeration technology) with strain sensitivity of 10-20 or better. This improvement comes both from cooling the antenna and from improvements in the transducer-amplifier readout. These improvements include the use of state of the art thin-film DC SQUID amplifiers. This can all be done with no basic changes in the antenna configuration. I should point out that the present antenna is made of aluminum. Recall that an important figure of merit in the antenna is the speed of sound. Use of an antenna material with a much higher sound speed is a future possibility that is highly desirable. The relevant scaling law comes from considering that the integrated antenna cross section $\sigma(f)$ df for the fundamental longitudinal mode of a cylindrical antenna can be written as (cGA/f) (v_s /c), where is the density of the antenna material and A is the cross sectional area of the cylinder. Clearly, at a fixed frequency, the energy sensitivity will improve as $(v_s)^3$ For aluminum this quantity is 3.6×10^{14} kg/sec. In beryllium the value is 4×10^{15} kg/sec. Thus, without even increasing the cross sectional area of the antenna (but increasing its length), a factor of 11 gain in energy sensitivity is available by using a different material for the antenna. There are also ceramic materials with very high sound speeds that may also be suitable.

It is clear that the detectors we are now operating are at the boundary of the sensitivity region in which we can reasonably expect to detect astrophysical sources of gravitational radiation. The next generation of detectors will push us much further into this region. I hope that I have been able to give you an idea of the scope and status of the two-pronged attack that is underway, using laser interferometry and resonant-mass detectors, on the problem of detecting gravitational radiation. The payoff of this long effort will be not only the detection of gravitational radiation but the opening of a new astronomical window on our universe.

MICHAEL E. FISHER:
Thank you very much. Very fascinating and a great challenge. We look forward to seeing the actual detection before too long.

PETER F. MICHELSON:
I must say that we need Fred Reines' detector on the air to do coincidences with - when the next supernova occurs. □□

In our next talk we move from the macroscopic on the very, very large scale, to things that can be done in the laboratory. It is a pleasure for me to introduce Albert Libchaber. He is a theorist experimentalist, in the sense that he has taken up some theories that challenge and fascinate condensed matter theorists. A theorist must actively put them to the test in seeing how well they actually work out. He was born in Paris, educated in France, and gained his first degree in 1956, and his Doctorate from the Ecole Normale. In 1983 he moved to the University of Chicago. That's where he is now actively doing work on chaos and turbulence, with some of the most beautiful results, recognized by the award of the Wolf prize in Physics in 1986. It's a pleasure to introduce Albert Libchaber.

□□

Experimental Gaze at Nonlinear Phenomena

Albert Libchaber

THE J. FRANCK AND E. FERMI INSTITUTES
THE UNIVERSITY OF CHICAGO

Chicago, Illinois 60637

I will talk about a field a little far away from physics, which up to recently, was within the bounds of mathematicians, metallurgists and engineers. I will try, therefore, to give you a feel of what the field is about. Essentially, we are trying to understand the evolution of patterns and rhythms in space and time, as some control parameters are varied. In this talk I will emphasize the historical evolution, and for convenience, describe experiments in which I was involved. Thus references will be incomplete and many beautiful experiments ignored.

Let me first talk about a great book. In 1917 D'arcy Wentworth Thompson wrote a beautiful book (1), called "*On Growth and Form*". Let me quote him: " *The waves of the sea, the little ripples on the shore, the sweeping curve of the sandy bay between the headlands, the outline of the hills, all these are so many riddles of form, so many problems of morphology, and all of them the physicist can more or less easily read and adequately solve*". "*Nor is it otherwise with the material forms of living things. Their problems of form are in the first instance mathematical problems, their problems of growth are essentially physical problems. There is no branch of mathematics, however abstract, which may not some day be applied to phenomena of the real world*". In his book D'arcy Thompson developed a program to show that in the living world, growth and form were at least not in contradiction with physical laws, and applied various branches of mathematics to look for the underlying unity of many processes.

This direct interplay between mathematics and natural phenomena is one of the characteristics of the field of nonlinear physics. We are often faced with the following problem: we know the equations related to a physical situation, but their intrinsic nonlinear structure does not allow us to solve them. For example in fluid flows we believe that the Navier-Stokes momentum transport equation and the Fourier heat equation represent the state of our flow.

But in general we cannot solve those equations, because of the nonlinear advection terms. What we try to do, and we sometimes succeed, is to write a simpler nonlinear equation, whose solutions would exactly mimic the results of our observations. In a way we look for the generic behavior of nonlinear phenomena.

This idea is very well expressed in the last paper of A. M. Turing (2) in 1952, called "*The Chemical Basis of Morphogenesis*", which is a study of coupled reaction diffusion equations, which are supposed to lead to spatial structures reminiscent of living organisms. He writes, " *A mathematical model of the growing embryo is described. This model is a simplification and an idealization, and consequently a falsification. It is to be hoped that the features retained are those of greatest importance in the present state of knowledge.*" So we observe out of equilibrium natural phenomena. We look at the development of patterns in space, and rhythms in time, and then we try to develop some simple modeling to explain our observations.

1. THE BIFURCATION TO A LIMIT CYCLE

Now, in order to give you a feel for the aesthetics of some of our problems, I will project a few figures. Instead of looking directly at natural phenomena, let me try to describe a controlled experiment like what Bechhoffer and Oswald (3) did, looking at the evolution of a pattern in space. They took a liquid crystal placed in sandwich between two microscope slides, thickness about 10 μm. The sample was placed in a temperature gradient, such that at the higher temperature it is an isotropic liquid crystal and at the lower temperature a nematic liquid crystal. The nematic-isotropic transition is at 40.5 °C. The inner glass surfaces were treated with silane, in order to orient the nematic phase perpendicular to the glass plates. The interface between the two phases is a flat front, as can be seen in Fig. 1a .

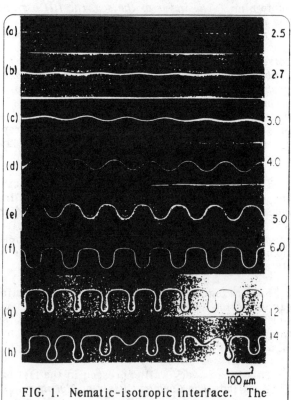

The experiment consists of moving the sample at a constant velocity from the hot side to the cold, while holding the ovens fixed in temperature and position (4). At a critical velocity of 2.7 μm/s an instability develops (Fig. 1b). The interface solution bifurcates from a flat solution to a wavy one. The amplitude of the wavy state being as small as we want, we have a supercritical bifurcation, analogous to a second order phase transition. The relevant physical mechanism is associated with the Mullin Sekerka instability (5), which describes pattern formation at onset in growing solid-liquid interfaces. Above onset various non-

FIG. 1. Nematic-isotropic interface. The nematic is at the bottom. Numbers indicate velocities in μm/s.

linear effects are observed.

Above 5 µm/s higher harmonics develop, leading to grooves where impurity atoms concentrate (Fig. 1f), and finally around 14 µm/s the grooves pinch off leaving behind bubbles of the isotropic phase with impurities. This experiment shows the simplest pattern evolution, starting from a flat interface state. We can naively speak of a Hopf bifurcation in space.

FIG. 2.   Cascade of triangles caused by the detachment of a defect line.
In the triangles the nematic orientation is planar, outside
homeotropic.   (See Libchaber page in color insert.)

So this is an example of the very general problem of the marginal stability of an interface.   But, in performing the experiment, Oswald and Bechhoffer found an interesting dynamical phenomenon connected to more traditional solid state physics.   Because of the plates' treatment the meniscus between the two phases is inverted, and for topological reasons a -1/2 defect line (6) runs along the interface.   A dust particle passing through the moving interface pins the defect line which then stretches out of the interface, and creates a triangle as seen in Fig. 2.   In fact in this figure you can see the effects of many dust particles.   Now, depending on the relaxing velocity of the line as compared to the velocity of the moving interface the effect will or will not be there.   Now, can a triangle split into two?   Can two adjacent triangles melt?   All these questions lead to an interesting evolution of the shapes of the resulting pattern.   This is an example of a development of a pattern in space.

"Development" is the important word; what we are focusing on is the evolution from order to disorder, as we change a control parameter, like the velocity of the front in the preceeding experiment.   To use Thom terminology (7), we look for the general unfolding around a transition.

FIG.   3.   Thermal convection in a large aspect ratio cell of mercury.
Horizontal magnetic field parallel to the general orientation of the
rolls.   The Ra number and the field are identical from (a) to (f).
Pictures are taken with about 10 sec. interval.   Visualization by
a cholesteric liquid crystal sheet.   Blue is warm, white is cold.
(See the Libchaber page in the color insert.)

Let me now give you an example of a development of a rhythm in   time.
Let us turn to Fig. 3.   It is a view of thermal convection in a mercury cell. As will be explained later, you get thermal convection by heating a fluid from below.   In this experiment (8), Joel Stavans used mercury, an electrically conducting fluid, and also he subjected the fluid to a horizontal D. C. magnetic field.   Now, mercury is opaque to light, so in order to visualize the pattern of ascending hot fluid and descending cold fluid, he placed a cholesteric liquid crystal sheet behind a sapphire plate.   The plate and the sheet represent the top boundary of the cell, and what you see in Fig. 3 is the pattern of convection in a cylindrical cell of about 10 cm diameter, and about 1 cm depth.   Blue regions indicate the arrival of hot fluid, and white ones the departure of cold fluid.   Figure 3 shows the evolution of the pattern for fixed temperature difference and fixed magnetic field value; it is a

pure oscillation in time, lasting a minute from fig. 3a to Fig. 3f, and the process repeats itself. The wavelength of the pattern is about twice the depth of the mercury cell, it is a pattern in space. Keeping all the control parameters fixed, this pattern will oscillate in time, the convective rolls will break and reconnect, shifting the whole pattern by one wavelength. In Fig. 3b you see the rolls' deformation, which will start to pinch (3c), then break (3d) and finally reconnect (3f), essentialy going back to the original pattern but with the rolls having advanced by one wavelength. The oscillation can last for ever, with a very well defined period of about a minute. So, just applying a DC temperature difference and a DC horizontal magnetic field, you can create a space pattern with a very precise time clock. The explanation, of course, is not simple. We have created a frustrated state, where the lateral boundary conditions impose that the rolls should be perpendicular to the boundary, and the horizontal magnetic field tends to align the rolls parallel to it. Those two constraints lead to the dynamics. Now the theoretical problem of rolls breaking and reconnecting is a very complex and unsolved problem, but the resulting dynamics is a simple limit cycle!

FIG. 4. Small aspect ratio mercury cell; (a) stationary convection; (b) the oscillatory state. (See the Libchaber page in the color insert.)

To simplify the dynamics, Claude Laroche took a small cell such that a small number of rolls can be accomodated (9), it is a rectangular one with 4 convective rolls, as visualized in Fig. 4a. It is what is called a small aspect ratio experiment, where space evolution like rolls breaking or dislocations cannot develop in the limited space (depth 5 mm, rectangle dimension about 5 mm → 2 cm). What happens if you increase the temperature difference is that the rolls start to oscillate back and forth towards the diagonal of the rectangle as shown in the snapshot of Fig. 4b. This oscillation has a period of about 2 seconds, and is related to the oscillatory instability of convection, introduced by F. Busse (10). This oscillation, or limit cycle, is the beginning of our construction of a chaotic state, for larger temperature differences , and even of developed turbulence!

In an experiment like thermal convection, as you increase the heat flux, you will reach a chaotic state and eventually a turbulent one, and this is what I want to tell you about, and also show you that it all happens in an ordered fashion through well defined change of states or bifurcations, if you work with a small aspect ratio cell. But before doing that I must introduce to you thermal convection.

Let us look at the sketch in Fig. 5, a fluid layer between two plates at temperature T and $T+\Delta T$, separated by a distance L. In the bottom the fluid is warmer, so due to the expansion coefficient α we have an upward buoyancy force $\alpha g\Delta T$ and thus an ascending time $(L/g\alpha\Delta T)^{1/2}$. This motion is opposed by a viscous drag force proportional to the kinematic viscosity ν (dimension L^2/T) and of characteristic time L^2/ν. Also, as the warm fluid goes up, heat diffuses away, in proportion to the heat diffusivity κ(dimension L^2/T) and with a characteristic time L^2/κ. A combination of those three times defines a dimensionless number called the Rayleigh number Ra, as was calculated by Lord Rayleigh in 1917 (11). For a critical value of Ra we will have a transition from heat diffusion to heat convection where the fluid starts to move, upward for the warm fluid and downward for the cold fluid. The value of Ra at the onset is not of order one, but of the order 10^3, because the fluid velocity has to go to zero at the top and bottom plates.

2. ROUTES TO CHAOS

So what we shall do in the remainder of this talk is increase the Ra number and observe what happens. It will all depend on the fluid you take. The fluid is characterized by another dimensionless number, the Prandtl number $Pr = v/\kappa$. It physically represents the ratio between two diffusion velocities, for momentum and for energy. We will be interested

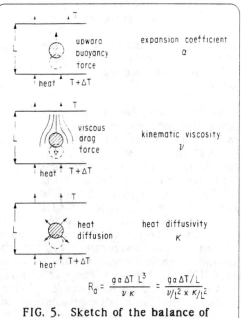

$$R_a = \frac{g\alpha \Delta T\, L^3}{\nu\, \kappa} = \frac{g\alpha \Delta T / L}{\nu/L^2 \times \kappa/L^2}$$

FIG. 5. Sketch of the balance of forces at the onset of convection.

here in low Pr fluids. For a gas, or for helium, the same mechanisms lead to momentum diffusion or energy diffusion, so Pr is about 1. For a liquid metal, energy transport is by the free electron gas, so Pr is the ratio between a sound velocity (momentum transfer) and a Fermi velocity, thus Pr is of the order of 10^{-2} to 10^{-3}. For a plasma, like the surface of the sun, the energy transfer is radiative, so Pr is a ratio between a sound velocity and the velocity of light, of the order of 10^{-6}. For viscous fluids, like oil, the Pr number is large. For our low Pr fluids, above the onset of convection, a bifurcation to an oscillatory state, called the oscillatory instability (10), shown in Fig. 4b, leads to a new state with a limit cycle. If we would then increase the Ra number, a second oscillation will appear, which in turn will lead to a chaotic state. To control precisely how it all happens, we want to force a second mode externally.

Fig. 6 is a skeleton of the next experiment I would like to describe, done by Joel Stavans and James Glazier (12). The fluid is mercury, the cell of aspect ratio two (2 convective rolls), the Ra number such that the fluid oscillates in velocity and temperature with a period of about 2 seconds, and a horizontal magnetic field of about 200 Gauss applied. We excite a second mode by applying an AC current sheet vertically, which induces an AC flow, in the presence of the magnetic field. We can control the amplitude and the frequency of this mode, by monitoring the current amplitude and frequency. We are thus left with two modes: one free mode resulting from the convective state, and a second one generated by electromagnetic forces, which we can control. How do we measure all that? Well, we have a bolometer sensitive to the local temperature, placed in the bottom plate of the cell (Fig. 6). In fact, it could be placed anywhere, the result would be the

same. So we start with two frequencies and our fluid has nonlinear properties. The Navier-
Stokes and Fourier equations in-
volve terms like $(v \cdot \nabla) v$ and $(v \cdot \nabla) T$ which will couple the two
modes.

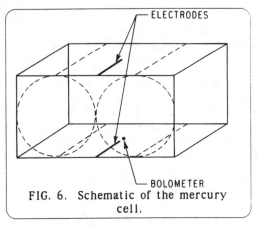

FIG. 6. Schematic of the mercury
cell.

We are now in the realm of the
theory of dynamical systems , introduced by Henri Poincaré (13)
in 1899 in "*Les Méthodes Nouvelles de la Mecanique Céleste*", and
a good modern presentation of the
field is to be found in the books
(14) of Vladimir Arnold. I will
only describe briefly the mathematics needed to explain the observations,
as we sweep the forcing frequency and amplitude. Surprisingly, the ratio
between our two frequencies is important, or more precisely whether it is a
rational or an irrational number! Let me remind you what Leon Brillouin
wrote (15) in 1959. "*The mathematician takes great care to distinguish
rational and irrational numbers. The physicist never meets an irrational
number*". Well, in this experiment, we come very close to it.

When you have a signal with two frequencies, you can give a representation of it in a phase space as shown in Fig. 7. It is a torus and the phases
of the frequencies correspond in the figure to the angles ϕ_1 and ϕ_2. The
cross-section of the torus, for $\phi_2 = 0$ for example, represents a Poincaré section, it is the limit cycle of the oscillator of frequency f_1 . If the ratio between the values of the two frequencies (the winding number) is a rational
number, the trajectory on the torus is a closed curve as shown in Fig. 7, and
the cross section a series of points. You have a locked state, where the free
oscillation is slaved by the forced one. If the winding number is an irrational, then your trajectory fills the torus ergodically, and the cross section is
the whole closed curve $\phi_2 = 0$. Well, without going into the details of how
we do it technically (16), we can reconstruct the Poincaré cross section of
our experimental signal. The irrational number chosen was the golden
mean, GM=$(\sqrt{5} - 1)/2$. In continuous fraction representation it reads:

$$GM = \langle 1, 1, 1, 1 . . . . . \rangle \quad \text{or:}$$
$$\text{Golden Mean} = 1/ [1 + (1/1 + (1/. . .))]$$

whose rational approximants are 1/2, 2/3, 3/5, 5/8, 8/13, 13/21, etc. In fact,
in the experiment, we just do that , slowly moving the forcing frequency to
all the approximants up to where our experimental noise allows us to go,
about 144/233 or one step more. When we are there, we increase the forcing amplitude and start again, until we reach a critical value for the amplitude, where the signal becomes noisy, in fact chaotic. This is very tedious
work and Stavans and Glazier spent long nights on it. Figure 8 shows the
Poincaré cross section of the experimental data at this critical point and for
a golden mean value of the winding number. If you are close by, at a rational approximant like 8/13, the cross section is as shown in Fig. 9a , where

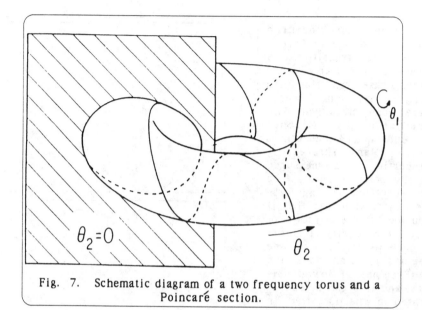

Fig. 7. Schematic diagram of a two frequency torus and a Poincaré section.

you can count 13 points: it is an 8/13 phase locked state.

If you prefer a less abstract representation, look at the temperature recording at a function of time, shown in Fig. 10. You will find a periodic oscillation with groups of 8 oscillations of the natural frequency of the mercury cell. Going back to the cross section, as you do the experiment, you see points being traced one after the other, like an iteration. This is what the mathematical representation of the experiment will be, a nonlinear iteration on a circle, called the circle map (17). If the winding number is an irrational, you find a critical value for the amplitude of the nonlinear term, where the trajectories become chaotic, as in the experiment. If you have a rational winding number, as you increase the amplitude you get a period doubling cascade developing, each point gives two points and so on.

For example, in the experiment, Fig. 9b shows that two period doubling bi-

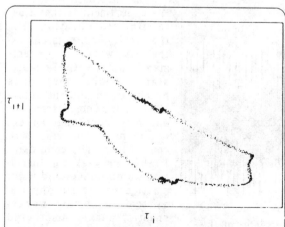

FIG. 8. Experimental critical attractor for a golden mean winding number.

furcations have occured and a third one is barely visible. The transition to a chaotic state is like a second order phase transition with critical exponants, which have been measured, and predicted by Mitchel Feigenbaum (18) and others (19). Many interesting properties can be measured at the critical point, like the spectrum of singularities. It just means that if you look at the attractor of Fig. 8 the density of points is not uniform. You can see regions of high densities and regions of low densities, and this can be represented in what is called, in the jargon, the spectrum of singularities or the multifractal representation or the $f(\alpha)$ function. Figure 11 shows the $f(\alpha)$ function for quasiperiodicity on the right and for period doubling on the left. The points show the results of the analysis from the mercury experiment.

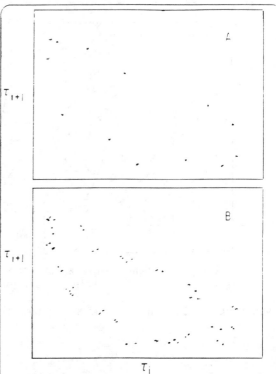

Fig. 9. Experimental Poincaré section in the 8/13 locked state. (A) a pure 8/13 state; (B) a 2^3 x (8/13) state.

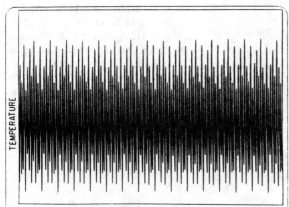

FIG. 10. Experimental time series of the local temperature for an 8/13 state in the mercury experiment.

The interesting result of the experiment is that a deterministic equation, like the circle map, mimicks extremely well the experiment, up to the chaotic state. There the torus breaks down and the analysis is more cumbersome, but one may hope that an experimental analysis will come (20). The conclusion is that for weak nonlinearities we do understand how a small aspect ratio physical system becomes weakly chaotic, weakly disordered, weakly turbulent, in a deterministic way. But of

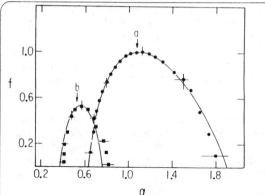

Fig. 11. f(α) function for quasiperiodicity (a), and for period doubling (b), at the
critical value. Points are the results from
the mercury experiment.

course we all know that the
atmosphere of Jupiter , or
the Earth, or the surface of
the sun, have much stron-
ger, much more well developed states of turbulence.
Can we understand the
transition from chaos to
turbulence, from a deter-
ministic approach to a statistical one? I would like to
describe what we are doing
now at the University of
Chicago in this context. It
is very simple to reach a
turbulent state, you just
have to increase a lot more
the heat flux. First, of
course , you have to in-
crease the size of the box, as the Rayleigh number goes as the size cubed.
Years ago, I was studying thermal convection in mm^3 of liquid helium,
then we went to cm^3 of mercury, and now I am telling you of an experiment where the characteristic size is about 10 cm. Also the fluid has
changed: helium gas at a temperature of about 5° K. This is the best situation to reach a very high Ra number, as shown by Threlfall (21). (Ra $\approx 10^{14}$
at the critical point).

3. FROM CHAOS TO TURBULENCE

Let me quote first Lev Landau from his paper "*On the Problem of Turbulence*", one of the important ones in this field (22). " *Although turbulent
motion has been extensively discussed in literature from different points of
view, the very essence of this phenomenon is still lacking sufficient clear-
ness. In the author's opinion, the problem may appear in a new light if the
process of initiation of turbulence is examined thoroughly.*" This is what
we did before, and we do it again with our liter of helium gas. The idea is to
start at very low pressure, perform your experiment, and then repeat at
different pressures up to the critical point of helium. You can easily
change the density by a factor of 10^5, and thus Ra by 10^{10}. Thus , starting
with a rarefied gas, Heslot and Castaing (23) could study convection at small
Ra number and then, opening more and more a low temperature valve ,
they could reach turbulent states at very high Ra number.

The inset of Fig. 12 is a sketch of the experimental cell, with the position
of the bolometers used to measure the local temperature fluctuations. The
plot represents the effective heat conductivity, called the Nusselt number,
as a function of the Ra number. Various transitions are labeled. At low Ra
number, they observe the routes to a chaotic state, but then for Ra > 2 x10^5,
two states of turbulence are described and related to a stable and an unstable boundary layer.

Why is that interesting? It is interesting in the following sense: The study of chaos is the game you play with the mathematics of time . You freeze space, small aspect ratio, and you play with oscillators in a nonlinear medium. What happens when you reach a turbulent state is that space comes in through a boundary layer formation, you have a stratification and different regions play a different dynamical game. For example, the boundary layer can be unstable to wave formation, as we shall see later. Let me introduce the boundary layer in thermal convection. Fluid mechanics is a beautiful testing ground for dimensional analysis, a concept introduced by Fourier in his book (24) " *Théorie Analytique de la Chaleur*". Well, I told you before that the Ra number is dimensionless, and scales with the cube of the height of the box, so it has to scale with another length cubed. This is the boundary layer thickness δ:

$$Ra = L^3/\delta^3$$

$$L/\delta = Ra^{1/3}$$

This is true within a scaling constant of the order of 10^{-1}, meaning that the onset of convection is for Ra ≈ 10^3. So in this simple argument, we have two length scales, a thickness δ near top and bottom plate where heat is just diffusing, and a center turbulent region. The last turbulent state in Fig. 12, called hard turbulence, is associated with an instability of the boundary layer, and strong emission of thermal plumes. To show you how different the temperature recordings are, as compared to the locked state shown in Fig. 10, look at the temperature recordings in Fig. 13, (a) for the center bolometer and (b) for the boundary layer one, in the hard turbulence regime.

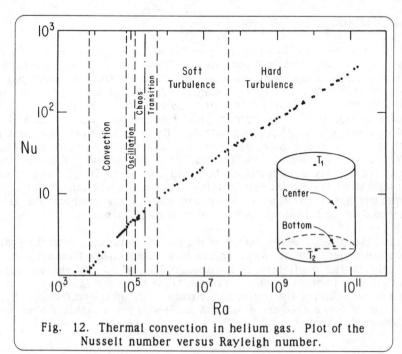

Fig. 12. Thermal convection in helium gas. Plot of the Nusselt number versus Rayleigh number.

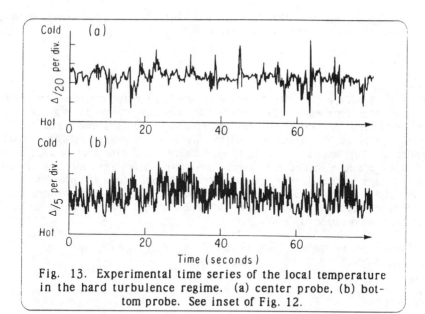

Fig. 13. Experimental time series of the local temperature in the hard turbulence regime. (a) center probe, (b) bottom probe. See inset of Fig. 12.

In the center probe, large and intermittent temperature spikes are present, indicating the arrival of thermal plumes.

I will not describe in more detail this experiment, which is in progress, but end up with a beautiful visualization that Steve Gross and Giovani Zocchi have just been able to obtain from the instability of the boundary layer at the transition from soft to hard turbulence (25). They took a cube of water, size 20 cm, heated from below, Ra number about 10^8 , and they shine a laser sheet at glazing incidence to the bottom plate. There is an onset for wave instabilities and as you increase the heat flux you can observe all of the various boundary layer wave states, isolated waves, interacting waves, breaking waves! This is hard turbulence and the boundary layers are playing a game different from the bulk of the flow. Fig. 14 for example shows breaking waves of the boundary layer.

Fig. 14. Visualization of boundary layer waves in thermal convection in water, Ra -10^9. (See Libchaber page in the color insert.)

4. CONCLUSION

Starting from a flat interface bifurcating to a wavy one, our journey took us up to turbulence, a bit too much for a simple talk. I have tried to show you various aspects of our subject, how lively it is, but also how classical it is, essentialy 19th century physics. The broad effort, in the last ten years, to understand the chaotic state is not described in this presentation, and I will refer to two good review books on the field (26). As I like historical references let me quote, in conclusion, Henri Poincare who at the end of the 19th century had a clear vision of the complexity of dynamics. Studying the intersection between homocline trajectories, he explains that the picture is so complicated that he cannot draw it. It is a good exercise to try to translate from the French language:

"*Que l'on cherche à se representer la figure formée par ces deux courbes et leurs intersections en nombre infini dont chacune correspond à une solution doublement asymptotique, ces intersections forment une sorte de treillis, de tissu, de reseau à mailles infiniment serrées;chacune des deux courbes ne doit jamais se recouper elle-même, mais elle doit se replier sur elle-même d'une manière très complexe pour venir recouper une infinité de fois toutes les mailles du reseau. On sera frappé de la complexité de cette figure, que je ne cherche même pas à tracer. Rien n'est plus propre à nous donner une idée de la complication du problème des trois corps et en géneral de tous les problèmes de dynamique . "*

H. Poincaré, *Les Méthodes Nouvelles de la Mécanique Céleste,* Gauthier-Villars, 1899, Vol. 3 , page 389.

REFERENCES:

(1) D'Arcy Wentworth Thompson, "*On growth and form*", Cambridge Un. Press (1917).
(2) A. M. Turing "*The chemical basis of morphogenesis*", Proc. Roy. Soc. , Vol. **237**. B 641, page 37 , (1952)
(3) P. Oswald, J. Bechhoefer, and A. Libchaber, "*Instabilities of a moving Nematic-Isotropic interface*", Phys. Rev. Let. **58** , 2318, (1987).
(4) K. A. Jackson and J. D. Hunt, Trans. Metall. Soc. AIME , **236**, 1929 (1966).
(5) W. W. Mullins and R. F. Sekerka , J. Appl. Phys. **35** , 444 (1965).
(6) P. G. de Gennes , *The Physics of liquid crystals,* Oxford (1974).
(7) T. Poston and I. Stewart , *Catastrophe theory and its applications,* Pitman (1978).
(8) J. Stavans and A. Libchaber " *Thermal convection in a small Prandtl number fluid* , Preprint.
(9) S. Fauve , C. Laroche and A. Libchaber, J. Physique Lettres (Paris), **42**, 455 (1981).
(10) F. H. Busse "*Non-linear properties of thermal convection*" Rep. Progr. Phys. **41**, 1929 (1978).
(11) Rayleigh (Lord) Phil. Mag **32**, 529 (1916).
(12) J. Stavans, F. Heslot and A. Libchaber, "*Fixed winding number and the Quasiperiodic route to Chaos in a convective fluid*" Phys. Rev. Lett. **55**, 596 (1985);
M. H. Jensen, L. P. Kadanoff, A. Libchaber, I. Procaccia and J. Stavans, "*Global universality at the onset of chaos:Results of a forced Rayleigh-Benard experiment*",
Phys. Rev. Lett. **55**, 2798 (1985);
J. A. Glazier, M. H. Jensen, A. Libchaber and J. Stavans, "*The structure of Arnold tongues and the f(α) spectrum for period*

doubling:experimental results" ,Phys. Rev. A. **29** , 811 (1984).

(13) H. Poincaré, *Les méthodes nouvelles de le mécanique céleste* ,
Gauthier-Villars, Paris (1899).

(14) V. Arnold *Les méthodes mathematiques de la mécanique classique*,
Mir, Moscow (1976);
V. Arnold *Chapitres supplementaires de la théorie des équations
differentielles ordinaires*, Mir, Moscow (1980);
V. Arnold, A. Varchenko and S. Goussein-Zade,
Singularités des applications differentiables , Mir, Moscow (1986).

(15) L. Brillouin, *Vie matière et observation*,
Albin Michel, Paris (1959), page 164.

(16) A. P. Fein, M. S. Heutmaker and J. P. Gollub *"Scaling at the
transition from quasiperiodicity to chaos in a hydrodynamic
system*" Phys. Scr. **T9**, 79 (1985);
J. Stavans, S. Thomae and A. Libchaber *"Experimental study of the
attractor of a driven Rayleigh Benard system*",
in *Dimensions and entropies in chaotic systems*,
Springer Verlag (1986), page 207;
P. Berge, Y. Pomeau and C. Vidal, *L'ordre dans le chaos:
vers une approche deterministe de la turbulence*,
Paris, Hermann (1984).

(17) M. H. Jensen, P. Bak and T. Bohr *"Transition to chaos by interaction
of resonances in dissipative systems. I. circle maps*"
Phys. Rev. A , **30** , 1960 (1984);
P. Bak *"The devil's staircase"*, Physics Today , **39**, 38 (1986).

(18) M. J. Feigenbaum *"Universal behavior in nonlinear systems*"
Physica **7D**, 16 (1983).

(19) S. Ostlund, D. Rand, J. Sethna and E. Siggia *"Universal properties
of the transition from quasi periodicity to chaos in dissipative
systems*", Physica **8D** , 303 (1983);
S. J. Shenker *"Scaling behavior in a map of a circle onto itself:
empirical results*" Physica **5D** , 405 (1982);
M. J. Feigenbaum, L. P. Kadanoff and S. Shenker Physica **5D**, 370
(1982).

(20) G. Gunaratne M. H. Jensen and I. Procaccia *"Universal strange
attractors on wrinkled tori*" Nonlinearity, Vol. 1 (1988).

(21) D. C. Threlfall, Journ. Fluid Mech. **67**, 17 (1975).

(22) *Collected papers of L. D. Landau*,, D. TerHaar ed., Pergamon (1965)
p. 387.

(23) F. Heslot, B. Castaing and A. Libchaber *"Transitions to turbulence
in helium gas*" Phys. Rev. A **36**, 5870 (1987).

(24) M. Fourier, *Théorie analytique de la chaleur*, Paris,
Firmin Didot (1822) p. 154.

(25) S. Gross, G. Zucchi and A. Libchaber *"Ondes et plumes de couche
limite thermique*" preprint.

(26) P. Cvitanovic, *Universality in chaos*, Bristol, Adam Hilger (1984);
H. Bai-Lin *Chaos*, World Scientific, Singapore (1984);
P. Collart and J. P. Eckmann, *Iterated maps on the interval as
dynamical systems* ,Boston, Birkhauser (1980).

A complete review of quasiperiodicity with extensive references :

J. A. Glazier and A. Libchaber "*Quasiperiodicity and dynamical systems: an experimentalist's view*". To appear in I. E. E. E. transaction on circuits and systems, August 1988.

KENNETH L. KOWALSKI:
I'd like to welcome you here again to Modern Physics in America this morning. We'd like to start immediately. We have a long but exciting program ahead of us.

The Chair for this morning's session is Mildred Dresselhaus. She is Institute Professor at M.I.T. She is also Director for the Center for Material Science and Engineering. She has had many accomplishments, among which was being elected President of the American Physical Society. Professor Dresselhaus.

MILDRED S. DRESSELHAUS:
I'm so happy to be here today to open up the second day of this wonderful symposium, and particularly to be here in Severance Hall. I've heard concerts from Severance Hall for many, many years. We get it on Public Broadcasting in Boston, but I've never been here. Within a twenty-four hour period I was able both to hear a concert and today I'm going to hear many wonderful lectures in this magnificent place. It's truly great to combine the arts and the sciences as did the two men that we honor with this celebration.

It's very fitting that we have a public display of science in this day and age, because of the great importance of science to the future of society, to our country, and to all of us here. Bringing science in public view is probably the most important thing that we practitioners in the field can do.

I'd like to take this opportunity to thank the Organizing Committee, the sponsors, the city of Cleveland, and everybody else that made this great event possible.

In the morning session we have three renaissance men who are going to tell us about the broad view of not only physics, but other fields that physics and physicists have touched on. Our first speaker, Dr. Ivar Giaever, started life -- well, he didn't start life as a mechanical engineer, but he started his professional life as a mechanical engineer in Norway, coming to the United States, first working for the General Electric Corporation as a mechanical engineer in problems of stress and heat transfer. Somewhere along the way in the '50's he was diverted from these practical goals to the field of physics.

As a very young man he had a wonderful idea that tunnelling in superconductors was important and we could learn something from that. It seemed very far out at the time, but it touched and affected the whole field of superconductivity bringing great excitement to it. Superconductivity was exciting already at that time because of BCS, but I think that Ivar added a great deal from the experimental point of view at that point. For this work he, at a very early age, I think he was probably less than thirty years old when he got the Buckley Prize, and then, soon thereafter, I guess it was about 1973, he got the Nobel Prize in physics for this great discovery. But by the time that all happened, Ivar was working in new frontiers. It was not physics of superconductivity, but it was immunology. He is no longer in this field, he's further and further into biology and today he's going to tell us, in his talk, how a physicist looks at biology. Ivar.

A Physicist's View of Biology

Ivar Giaever

GENERAL ELECTRIC RESEARCH AND DEVELOPMENT CENTER

Schenectady, New York 12301

I think I have set myself an impossible task this morning. I'm going to talk to you about biophysics. I had a Guggenheim in the late sixties and I went to Cambridge, England to study biophysics. When I told the people what I was going to do, invariably they said, "You poor man, there is no such thing." The more I have learned about biophysics, the more I tend to agree with the English that there is no such field as biophysics. At least it is not a very coherent field, say like nuclear physics or solid state physics. You really have to define biophysics in a very pragmatic way: biophysics is what physicists do when they do biology.

The way I'm going to explain biophysics to you is to just tell what I did when I started in biology. That's at least interesting to me, and I hope it's also interesting to you.

Now as you heard from the introduction, I did some tunnelling work. When I learned some biology in England, I came back to the General Electric Company and said "I'm a biologist now." But then you have a serious problem because you have to pick a problem to work on. That's what professors are good for if you're a graduate student as they tell you what problem to work on. But when you're on your own, you have to figure it out for yourself and that's a very difficult thing. What physicists tend to do when they enter biology is to pick a problem that requires techniques and skills that they are already familiar with.

Since I had dealt with electron tunnelling and I thought one way I could contribute to biology, was to make a much better microscope. You may not have thought about it, but we have wonderful ways of looking at very small things in nature using x-rays, like atoms in a crystal structure. We can look at larger things by using optical microscopes and electron microscopes, but the molecules which are important to biology we cannot look at very well. Those are protein molecules and DNA.

Now, the way I got started was by using the field-ion microscope invented by Mueller. This is a very simple microscope. As shown in Fig. 1, it has a tungsten needle which faces a television screen in a vacuum. Now, if you apply an electric field, electrons will tunnel out and follow a field line, bang into the television screen, excite the phosphor and be visible. Unfortunately, electrons also have a small random transverse velocity and therefore a point on the tip will correspond to a small area on the screen. So even though you have a magnification which is roughly the radius of the screen to the radius of the tip, the resolution is not good enough to see single atoms because of the electron smearing.

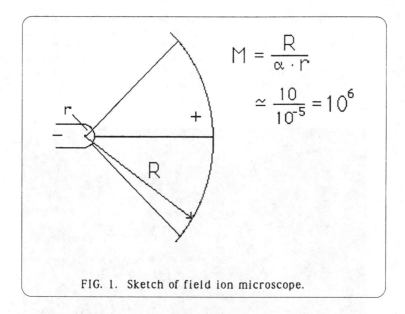

$$M = \frac{R}{\alpha \cdot r}$$

$$\approx \frac{10}{10^{-5}} = 10^6$$

FIG. 1. Sketch of field ion microscope.

Now Erwin Mueller did a very clever thing, he reversed the voltage. He had the screen being positive and the tip being negative. Then, of course, you get no electrons out. But, if you have a few atoms in the vacuum you have a dilute gas of hydrogen, say, hydrogen will go to the tip, ionize and then follow the field lines and bang into the screen. So this is the way Mueller did it. Now, if you like Mueller, you say this is a wonderful invention. Mueller was a somewhat controversial figure so if you didn't like Mueller, you say one day he reversed the voltage by mistake. Either way, the experiment works. Let me show you one of his pictures.

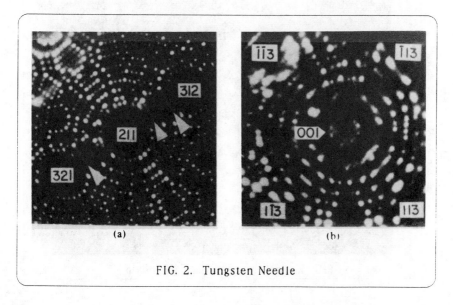

FIG. 2. Tungsten Needle

Figure 2 is a picture of a tungsten needle and here you really see single atoms very well. I'd like to remind people about these kind of pictures because today the scanning tunnelling microscope has become very popular. Mueller had already seen single atoms long before. There are many requirements to get the Nobel Prize. One requirement, unfortunately, is that you have to be alive, and Mueller died before his time came up.

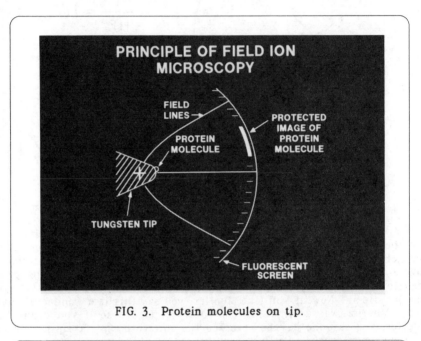

FIG. 3. Protein molecules on tip.

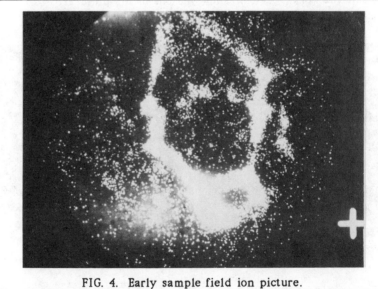

FIG. 4. Early sample field ion picture.

Now, the idea I had was not a very spectacular idea. It was simply to put protein molecules on the tip, as indicated in Fig. 3. The ions will still come to the screen, but the shadow of this protein molecule should be visible. So that was the basic idea I started with, but it didn't work very well. Actually, the kind of pictures we see is shown in Fig. 4.

I was doing this work with John Panitz in the Sandia Laboratories. These pictures are interesting to people who do field-ion microscopy, but they are not very interesting to people who do biology, because you really can't unravel this picture and go back to the structure of the protein molecule. So I got very discouraged with this. Even though this scheme sort of works, it doesn't work very well. What you really want to do is to try to help the biologist. One thing this got me into since I also worked with thin film, was I got worried whether protein would stick to a tungsten tip. So that was the problem I settled on next. Let me first of all tell you a little bit about protein, since most of you are physicists.

Protein is a common name for a class of molecules in your body. They have a high molecular weight, roughly 10^6 or so, but it varies for different kinds of proteins. For example, many hormones in your body are made of protein; practically speaking, all enzymes in your body are made of protein. Protein is a polymer molecule and the monomers are 20 different amino acids. The amino acids in different proteins are ordered in different ways, and this is what's called a primary structure. When a protein is folded up, it may consist of two or three chains and it will look schematically like that shown in Fig. 5. What I became very interested in, since I had dealt with these tungsten tips,

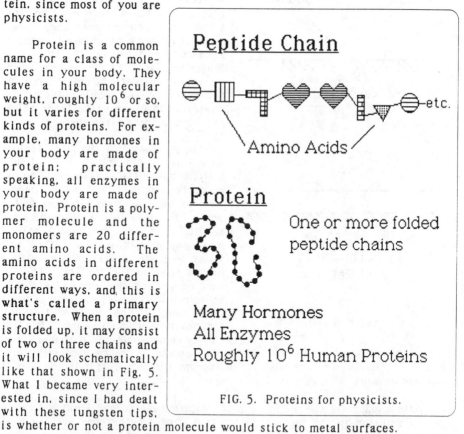

FIG. 5. Proteins for physicists.

is whether or not a protein molecule would stick to metal surfaces.

Now you think that you can go into literature and look that up, but you see biologists are not interested in metal surfaces, only physicists are. Nobody had studied adsorption of protein on metal surfaces (actually it turned out that A. Rothen had, but I couldn't find it at the time). So I decided that I

was going to do that. And to do this particular kind of work, I worked with
an ellipsometer, sketched in Fig. 6.

If you shine light onto a surface, the light is reflected. If you shine polarized light onto a surface, the light is reflected elliptically from the surface and from that you can calculate the index of refraction of a metal surface. Strangely enough, if you put a thin film on the surface, the ellipticity
of the light changes quite a bit. From that change you can get the index of
refraction and the film thickness.

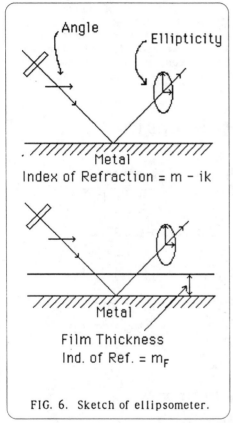

Angle

Ellipticity

Metal

Index of Refraction = m – ik

Metal

Film Thickness

Ind. of Ref. = m$_F$

FIG. 6. Sketch of ellipsometer.

The ellipsometer was invented by
Drude in the 1880's. The very surprising thing is that you can measure
film thickness to a fraction of an
Angstrom, to less than one Angstrom.
So, it's very sensitive. You recognize
that when I say a fraction of an Angstrom, that you measure the average
film thickness integrated over an
area corresponding to the wave
length of light. It's a very, very sensitive instrument. It's now used extensively in the semiconductor in-
dustry to measure oxide layer thicknesses and things like that.

Let me show you the first experiment I did in biology. In Fig. 7 I show
a little bucket of salt water. You can
actually use distilled water if you
want to, but then biologists get mad at
you. You have to use physiological
saline, otherwise they won't believe
what you do. Into a bucket of salt
water you put a glass slide with a
nickel film on the surface. Then at
time equals zero you drop a particular
protein into the bucket. This particular protein is known as BSA. It stands
for bovine serum albumen. It's the
most common protein in the blood of
a cow. It plays the same role in biology as the hydrogen atom plays in
physics. If it doesn't work with bovine serum albumen, forget it. So this is
what people work with as a model system.

I have a magnetic stirrer to keep the solution mixed and I put a drop of
protein solution in the bucket. I want to know whether the protein will
diffuse over and absorb on the nickel surface. While I do this experiment
the sample is mounted so that the light gets reflected and detected by the ellipsometer and I can tell whether the ellipticity changes. From this you
can calculate the film thickness.

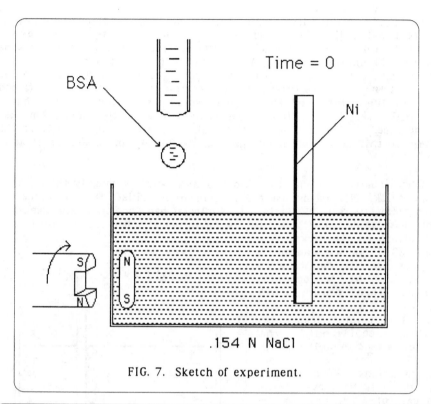

FIG. 7. Sketch of experiment.

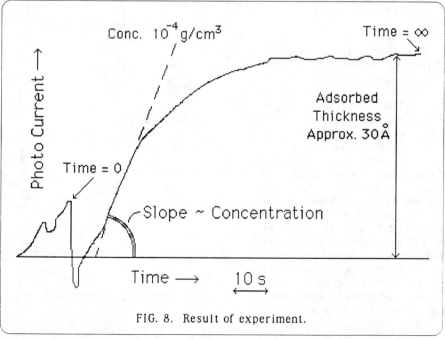

FIG. 8. Result of experiment.

The result is something like Fig. 8. The abcissa is time, the ordinate is photocurrent that is directly proportional to the film thickness. As soon as you
drop BSA into this little bucket, the photocurrent increases, which means
that the thickness increases. After roughly thirty seconds or so, or a
minute maybe, nothing more happens. Now, you see, there are some big
wiggles at the left. The reasons for those are purely experimental. (I had
to borrow the ellipsometer and I had to work in the pitch dark in a basement at General Electric. If you drop protein into a little bucket, you miss
all the time. So I had to hold a flashlight in my left hand, and it wasn't
steady. So this is my left hand shaking, it's purely an experimental artifact.)

What you learn is that the adsorbed BSA layer is roughly thirty Angstroms thick. Now let me use the experiment to explain the difference between biologists and physicists. If a biologist had done this experiment, he
would have said that if you drop bovine serum albumen into a solution, it

will absorb on the nickel surface at 25 °C, in one atmosphere
of pressure if you live in
Schenectady, NY, or something
like that. When I did this I took
the physicist's point of view. I
said: "The protein sticks to everything except itself." What
physicists tend to do is generalize. They just make enormous
generalizations. What biologists
tend to do is be very specific
and careful. The best way is
probably to do something in between.

Now, the next experiment I
did is in Fig. 9. Here is the bucket, but now I have protein on
the nickel surface, a layer of
BSA. Then I will take some
serum from the rabbit and drop
it into the bucket. [That's blood
from a rabbit from which all
the cells have been removed.]
In the rabbit serum there are
roughly a hundred thousand
different kinds of protein molecules like albumin, insulin, antibodies, all sorts of different
kinds of molecules. So, I ask the
ellipsometer: "Do any of those
kinds of molecules absorb?"
The ellipsometer said: "No, they
do not." (Therefore I don't even
have a slide of that because, you

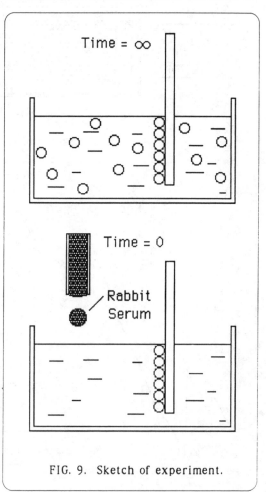

FIG. 9. Sketch of experiment.

know, if nothing happened then there's no point in making a slide for it!)

But I can extend the rule. The new rule says that protein sticks to everything except itself and any other protein. So that's a very simple rule. It's a rule of thumb, of course, it's not true for everything. There is one major exception. That comes when I deal with antibodies, when I deal with immunology. That's what I'm going to tell you about next.

The only way that I can learn something is to be specific. I don't understand when people talk in general terms. Let me be specific. Here in Fig. 10 is a person. When he gets exposed to a virus, I don't care what kind of virus it is, it can be chicken pox or polio, he gets sick. There's no question about that. Now, it's a strange thing when you get sick from a virus. You think that you can go to the medical profession and say, "please do something for me," but there's nothing they can do. That's a well kept secret by the medical profession, by the way.

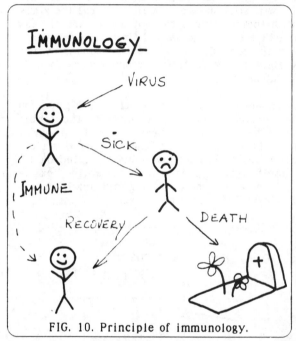

FIG. 10. Principle of immunology.

If you get a bacterial illness, you can take antibiotics but if you get a virus illness, there are no cures. Now when you get sick two things can happen. You can die in which case we don't have to worry about it. But, also, you know that most people recover and that is due to the immune system. The other remarkable thing is that the next time you get exposed to the same virus, you don't even get sick. You go directly to GO and collect two hundred dollars. If it weren't for the immune system, we wouldn't be here.

How can this system work? Well, it's very mysterious, very complicated really, but you can also make it simple. In simple terms the virus is a foreign particle in your body. The virus will get inside a cell and it will multiply itself and then the cell will burst. You now have a thousand viruses. Those thousand viruses go into a thousand cells and they will multiply and they will burst, and a million viruses will go into a million cells, and away you go. At the same time, you have something called B cells in your body. They also recognize the virus and they start making a protein molecule called an antibody. They make antibodies as fast as they can and the viruses multiply as fast as they can.

The purpose of the antibody is to attach itself to the virus. It specifically attaches to it by a chemical linking. When the antibody attaches to the virus, a class of white blood cells called macrophages recognize that this is a foreign object. The white blood cells eat up the virus and that's how you cure yourself. You see, it's a race between life and death. The viruses multiply. The B cells makes antibodies. If you make enough antibodies, it will tag all the viruses and you get well. If you don't make enough antibodies, the viruses win and you'll die.

When I grew up in Norway, a large number of people died from polio because they couldn't make antibodies fast enough. Now, of course, you get vaccinated. The way that works is you get injected an altered virus that cannot multiply, but the B-cells will still make antibodies towards the virus. Then when you get infected with the real virus the antibodies are already in your body. Therefore, you don't get sick the second time because the antibodies are there in sufficient numbers. So that's how the immune system works. It's very simple, very elegant. The details are much more complicated, of course, than I can talk about here.

Now I'm going to do the experiment over again, as sketched in Fig. 9. I'm going to leave the BSA on the surface. I'm going to drop in the rabbit serum as before, but this rabbit has previously been injected with BSA. When it's injected with BSA the rabbit doesn't know whether this is a sickness or not, but it makes antibodies to any foreign object, including the protein from a cow. So this rabbit's serum presumably has antibodies towards BSA in it. Now we can ask the ellipsometer: "Does anything from this rabbit serum absorb on the BSA layer?" The answer is yes.

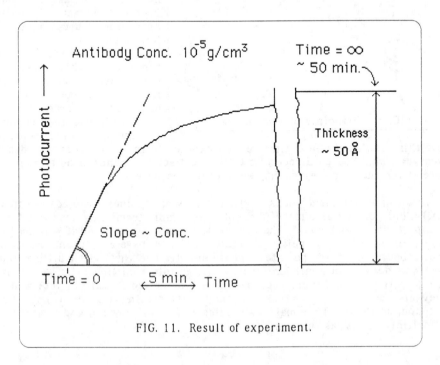

FIG. 11. Result of experiment.

Fig. 11 shows the results. The photocurrent increases. This takes roughly about an hour now because the concentration is lower. The new layer on
the surface is roughly 50 Angstroms thick. So you can use the ellipsometer
to find out whether you have antibodies in your system. This is sort of an
exciting figure, I think, because what you really see here is the result of a
chemical reaction as it goes on. The way a physicist looks at this is schematically shown in Fig. 12.

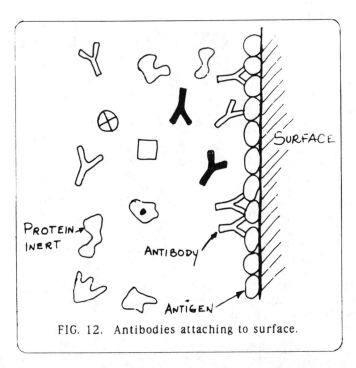

FIG. 12. Antibodies attaching to surface.

The antibodies are divalent. Here you see the antibody has attached and
you have some other antibodies (for chicken pox or something) but they
will not attach. Only the antibodies which have the right kind of recognition site will attach and start forming this double layer. That is what you
detect with the ellipsometer and that is what the doctors detect in a hospital,
but by using other methods.

Whether you like it or not, you carry your life history with you in your
blood. If you give me a drop of blood I'll tell you what you have been up to,
if you want me to. Now I became very excited about this, and I thought this
was a wonderful way of doing immunology, to detect all kinds of illnesses,
like syphilis and AIDS, and so forth. I didn't know much about hospitals at
that time. I have now worked in hospitals and I recognize that while ellip-
sometry is a good instrument in the hands of a physicist, it's not a very useful instrument in a clinical laboratory.

We believed in the simple rule that protein sticks to every surface except a surface covered with protein. No other protein sticks to it unless it is a specific antibody. So we invented another method, which is called the indium slide method. This is sort of a physicist's way of doing things.

You start out with a layer of glass or plastic as shown in Fig. 13. Onto this surface you evaporate indium. When you do that indium agglutinates into little particles and it does this just by itself. Now, when you take the indium slide and dip into a solution of protein, the protein will absorb on the indium particles. Remember, I said protein absorbs to everything; that includes indium particles. You remove the slide, rinse it, dry it, and you look at it. You can see the layer which is roughly between 20 and 30 Angstroms thick because the protein is adsorbed onto the metal particles.

Indium Slide Technique

← Glass or Plastic

← Evaporate Indium that Agglutinates into Particles

← Deposit Layer of Protein [or Langmuir Film] by Dipping into Solution

Layer is Visible!

FIG. 13. Indium slide technique.

FIG. 14. Photo of indium slide.

Now let me show you how it actually looks. Fig. 14 shows an indium slide. Here is a layer of protein. This is the BSA (you should know that because I wrote on it). You can clearly see the difference in contrast between the naked slide and the protein layer. Now this is not photography. It is a real slide in the projector. So, you're looking at the results of a real experiment, not some picture or anything like that. All you need is a Kodak projector or a lightbox to look at a layer of protein which is 20 Angstroms thick.

In Fig. 15 you can see the indium particles in an electron microscope picture. You can make them small, medium and larger. The experiment works best when the particle is roughly the wavelength of light. That's God's way of getting back at you; you can't make any approximation in this regime. If you're going to solve this problem, you've got to solve Maxwell's equations in all their glory. People have done that. Basically you have a small metallic particle and a thin dielectric layer changes the light scattering quite a bit. I have a small

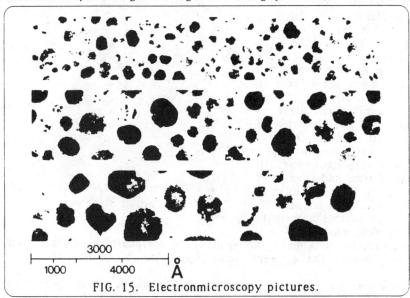

3000

1000   4000   Å

FIG. 15. Electronmicroscopy pictures.

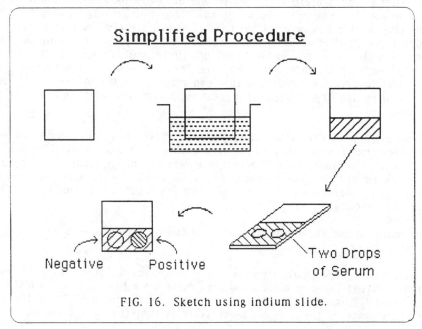

Simplified Procedure

Negative   Positive   Two Drops of Serum

FIG. 16. Sketch using indium slide.

particle with a dielectric layer: a layer of protein. This system would work with any kind of metal particles, indium happens to be easy.

Now, let me show you how you do immunology with this slide. The process is illustrated in Fig. 16. You take the slide with indium on it, dip it in a solution of protein, rinse, dry it, and you can see a layer of protein as I already showed you. You place the slide on the table and put two drops of blood on it: one with antibody and one with no antibody. You wait an appropriate amount of time and then wash it off, dry it, and you look at it. If you didn't have any specific antibody to this protein in your system, the layer will

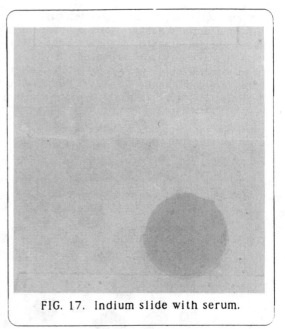

FIG. 17. Indium slide with serum.

not increase in thickness. If you did have antibodies, you'd get a double layer of protein. That's a very simple way of doing immunology.

I show you how it actually looks in Fig. 17. At the top is the naked slide and below is the protein layer. I had one drop of blood at the lower left, I swear to you, but you can't see it. It was there, but nothing adsorbed. At the right you can see the layer of antibody. You can clearly see this works well to detect antibodies. I have spent a large amount of work trying to make this into a clinical test. So far I have not succeeded. It's very difficult. It's easier to publish a paper in the *Physical Review* (you may not believe that) than it is to produce something that can make it in the marketplace. But the World Health Organization is a great supporter now, so I hope it still has a chance.

Immunology, as I've talked about, is really like chemistry. Here you have molecules that react, they're very predictable. It's a very nice field to be in, in a way, because it's reproducible if you do things right. Of course if you work in biology, the ultimate thing is to work with a living system. That's what I wanted to do. Then of course you have a choice of so many different systems.

You could choose an elephant, you could choose a flower, you could choose anything which is living, as in Fig. 18.

Amazingly enough all living systems are made up of cells. Everything which is living is made of cells; bricks make up a building, cells make up everything living. Now, an elephant is made up of roughly 10^{14} cells or so. The other amazing thing about these cells is that they are alive by them-

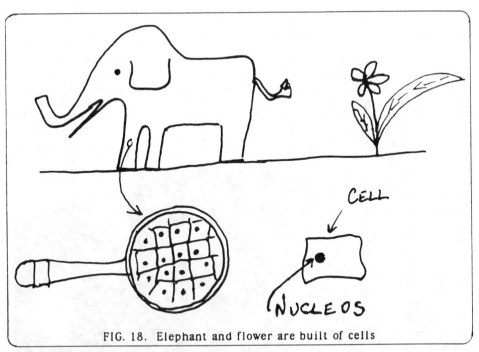

FIG. 18. Elephant and flower are built of cells

selves. You have 10^{14} cells in your body and all of these cells are independently alive. It's a mystery to me that when I wave at you with my hand,

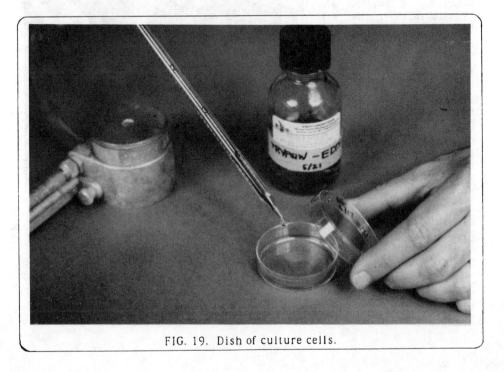

FIG. 19. Dish of culture cells.

that the cells can cooperate in this way. Hundreds of millions of cells in an ultimate cooperative phenomenon.

The interesting thing to me is that you can take these cells apart and grow them independently of an organism. That's the kind of work I'm doing now. You take little plastic dishes as in Fig. 19. You put a liquid into these plastic dishes, which the cells like to eat or drink, or whatever they do. These are dishes about 10 cm in diameter and you place the cells in there. Now, these mammalian cells are alive by themselves. They will have an independent life similar to a bacterium.

The only other thing you have to do is to keep the dishes in an incubator (as shown in Fig. 20) kept at body temperature, 37 °C. The incubator also has an enriched atmosphere of CO_2, as the buffer depends on the CO_2 concentration. In front you see Charlie Keese who has taught me all about cells. He has a "green thumb" when it comes to culturing of cells. It is more like an art than a science. I suspect that Charlie talks to the cells, but he vehemently denies it!

Let me show you a couple of pictures here. On the top of Fig. 21 there are lots of cancer cells and on the bottom are some normal cells. The normal cells grow in an orderly fashion but the cancer cells grow randomly. You see the difference between the cancer cells and the normal cells by the way they grow,

FIG. 20. C.R. Keese in action.

by the pattern they form. Now, believe it or not, if you have cancer and the doctor takes a biopsy, the only way he can tell you whether you have cancer or not is to look at the pattern the cells form. So, it's a completely subjective technique. There's no objective way of differentiating cancer cells from normal cells. Somebody looks at it and he says, "yes, those are cancer cells", or, "no, those are normal". But the doctor is forced to say yes or no. He has no choice in the matter. However, there is no objective way.

Cover of SCIENCE, 19 June 1981,
vol. 212, Chester Reather
FIG. 21.   Copyright 1981 by the AAAS.
(Cancer cells on top, normal on bottom)

One reason that people like me work in tissue culture is that I hope to find some powder, which I can spray on the cells. If they're cancerous they turn red, and if they are normal they turn green or something. I don't have such a powder, but that's the sort of thing we dream about when we do this kind of work.

Another very interesting thing about cancer cells and normal cells is that when these cells are grown in tissue culture they grow fat, then they divide and you have two cells. Cancer cells will continue to divide forever. They will go on day after day, year after year. They divide roughly every twelve, eighteen hours. Normal cells only divide roughly fifty times and then they die. Now why is that?

We all started from a single cell, from a fertilized egg. A chicken egg is a single cell. The cell divides and you have two cells, then four cells and so on. When a single cell has divided roughly fifty times, you're my age and you see what is happening to me. (From now on, it's downhill!) The only way I can live eternally is to get cancer. Now that's a contradiction in terms. Certainly if I got cancer, the cells that carry my DNA will go on forever (if somebody cared enough to keep them going in a laboratory) but my normal cells will not.

Now, let me try to explain to you specifically the things I'm interested in. Fig. 22 shows a blown-up picture of a cell: it's a spreading fibroblast. Indicated is the nucleus where the DNA is found. The cell normally is rounded in suspension. When it grows in tissue culture, it's attached to the surface and stretches out. So, this is not an equilibrium position. One very interesting thing is to measure the forces the cell exerts on the surface. The cells never grow directly on the plastic dishes, they always grow on a protein layer which is already there. I'm very much interested in this protein layer, because I already know about that from my study in immunology. I'm interested in how the cell attaches to the surface. There's a glue that holds the cell to the surface.

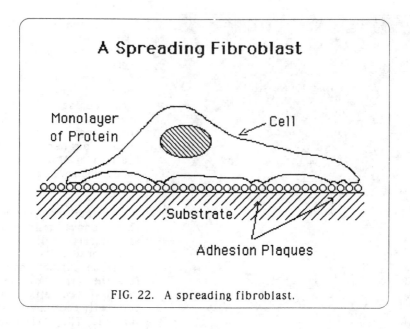

FIG. 22. A spreading fibroblast.

If you're unfortunate, say, and get cancer of the breast, nobody really dies from the cancer of the breast as such. You get lumps in the breast, but that is not dangerous. What is dangerous is that the cancer cells in the breast don't know they're breast cells; they let go and they go in the blood stream and settle in the brain, lungs or the bone marrow and kill you. So the reason cancer is very dangerous is that it metastasizes; the cells want to strike out on their own. They refuse to cooperate with the organism. They say: "I'm a breast cell, but I can do better in the lungs". But while doing better for a short time it kills you by growing in the lungs. Nobody really knows why a cancer cell metastasize. That means they have a different attachment mechanism than normal cells. A different glue

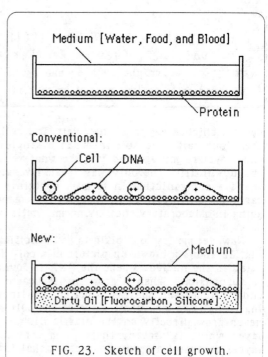

FIG. 23. Sketch of cell growth.

holds the cells to the surface. Thus the ultimate goal, which I am reaching for, is to understand the attachment mechanisms. I'll probably never get there, but that's what I'm trying to do at the present time.

When you normally grow cells in these dishes, as in Fig. 23, you have the tissue culture medium. Then you get a layer of protein on the surface and put in the cells which will first be spherical. Then they will attach to the dish, stretch out and grow. If they like what you feed them, they will get fat. They'll round up and divide and then you have two cells. That's basically the cycle of the cells in tissue culture.

People ask us how do we know that the cell attaches to a protein layer? How do we know the cell doesn't simply eat up the protein and attach to the plastic dish? What we do to answer this question is to put a liquid underneath. We use a fluorocarbon oil which is heavier than the water. So if you now have a liquid underneath, the cells will grow on the protein layer absorbed on the liquid surface. Then, of course, you'd know it can't remove the protein because if it did, it couldn't attach to a liquid. If it stretches out it has to have a substrate that can support a shear force. So it has to be growing on top of the protein layer.

I show you how this looks in Fig. 24. These are some normal cells growing on top of the liquid substrate we use. There are lots and lots of cells

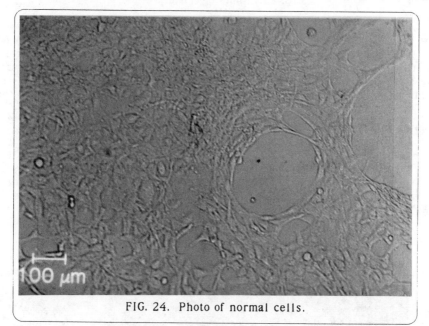

FIG. 24. Photo of normal cells.

here. You can see an opening. What happened was that the cells pulled too strongly on the protein layer and tore it; it relaxes back and it forms this rounded opening which we call a lake. So the cell was too strong for the protein layer.

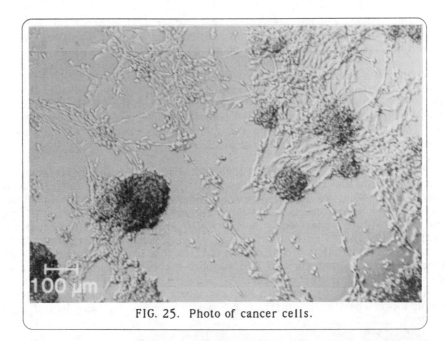

FIG. 25. Photo of cancer cells.

Now Fig. 25 shows the same surface with cancer cells. They grow in an entirely different pattern because they are so strong that when they stretch
out, the protein layer breaks immediately. They tend to grow in these wart-
like structures which is a very different pattern. In this work, I used some
commercial flurocarbon oil which contained impurities. That means that

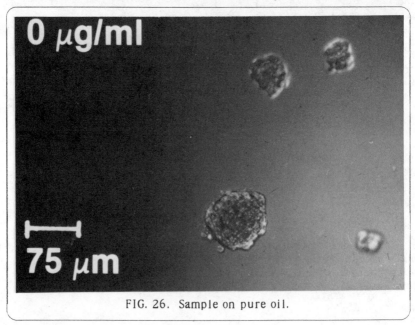

FIG. 26. Sample on pure oil.

the cells could attach very well. If you use very high grade flurocarbon oil, cells cannot attach at all.

This is shown in Fig. 26. If you have a very pure flurocarbon oil, the proteins on the flurocarbon oil then are like ping pong balls in a swimming pool. Every time you grab onto one it just moves. It moves among the others and they can't support any shear forces. If you put some impurity in (there are different impurities you can use) you can immediately strengthen this layer and the cell can grow.

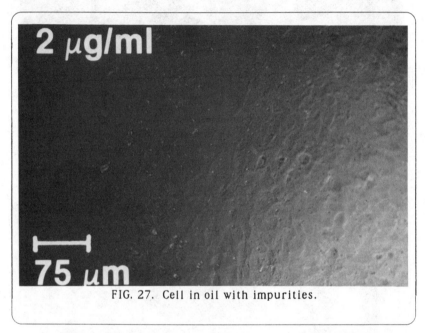

FIG. 27. Cell in oil with impurities.

Fig. 27 shows the effect of introducing a little bit of impurity. Now you have a complete layer of cells. You can actually regulate the strength of this protein layer by the amount of the impurities you put in. We have used that to measure the forces that cells put on the substrate.

We used a computer to measure how the cell can stretch out and how big they are compared to the strength of this protein layer. Such a computer result is shown in Fig. 28.

The way we measure the strength of the protein layer is to use a paddle wheel, as shown in Fig. 29. You can turn the torsion wire while the paddle wheel is stuck in the protein layer. When you turn it enough you find that the paddle wheel will release, and you can get the strength from that. Then you can correlate these two measurements, how far the the cells spread and how strong the protein layer was. Then you can get the average value for how much the cells pull on the surface.

One more thing I want to do is to show you that, since I work for General

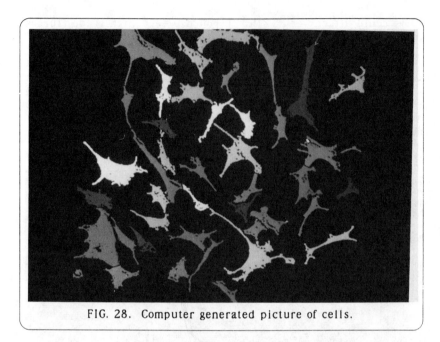

FIG. 28. Computer generated picture of cells.

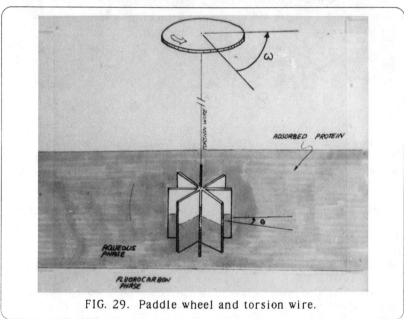

FIG. 29. Paddle wheel and torsion wire.

Electric which is a profit-making organization, you can turn this into a product.

We grow cells on an emulsion of fluorocarbon oil because if you want to grow cells in tissue culture, to produce products like fibronectins, growth hormones or enzymes, you need large surface areas. The way to get large surface areas in a small volume is to use small particles. Normally small plastic beads are used, but we can grow these cells on an emulsion of both fluorocarbon and silicon oil. A sample is shown in Fig. 30.

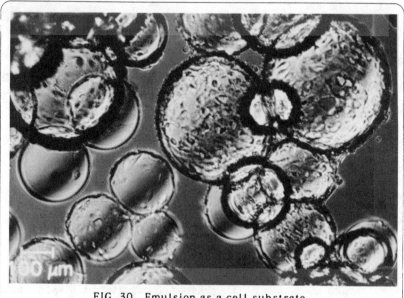

FIG. 30. Emulsion as a cell substrate.

Now let me talk to you a little bit about electrical and magnetic fields. Let me dispose of the magnetic fields first because that's very quick. Everybody loves to do biology in static magnetic fields (Fig. 31). *Science and Nature* have roughly one or two papers every year about the effect of magnetic fields on living things. I'm going to make a flat statement that there are no effects of static magnetic field on living things. I actually should be more careful and say there are no *irreproducible* effects of magnetic fields on living things. There's one exception however. The people in Woods Hole have discovered magnetotactic bacteria and these are bacteria which actually have a ferromagnet inside. Without the ferromagnet I have never been convinced that static magnetic fields have any effect on living things.

Now we can look at effects of the electrical field which is much more interesting. I started looking at this I because I am very much interested in electrical fields, as are all physicists. Do they effect the living state? The way I did the experiment was to place two electrodes in a tissue culture dish, put cells on the electrode, applied an electric field and observed what happened to the cells. This is sketched in Fig. 32. You can apply a small field, let it go for a while, then look at the cell, and the cells look happy. So you apply a larger field, you look at the cells, and the cells are still happy. So you get mad (you're in complete control) and you really crank up the field.

Then you look at the cells, and by gosh they're dying on one electrode. You say, oh, this is wonderful, electrical fields are really a bad thing for the cells.

I have a good friend at General Electric named Charlie Bean and he put me straight on this particular problem. He reminded me that when you have high fields you change the pH at the electrodes. You get all sorts of electrochemical reactions on the electrodes and the electrochemical reaction is what's killing the cells, not the electric field. Therefore, this is a "pseudo-effect" of electric fields. A lot of people fall

Magnetic Field Has No Effect on the Living State

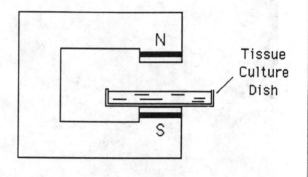

Tissue Culture Dish

Exception:
Magnetotactic Bacteria

FIG. 31. Sketch of magnetic field experiment.

Pseudo Effect

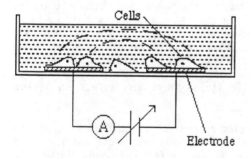

Cells

Electrode

Electrode Reactions
Change in pH, ion conc., etc.

FIG. 32.

into that trap. I'm not saying that electrical fields don't have an effect on cells. I'm saying it's very difficult to do reproducible experiments. The electric chair is one example where there definitely is an effect of electric fields on a living system.

So, I decided the experiment was really too difficult and I said I'm not going to do that anymore, but I'm going to see if cells can effect the electrical field. That's a very simple thing to do. Here in Fig. 33 you have the surface and you have an electrode. The cells come down rounded, attach and stretch out. The current has to flow in a different pattern. If you can detect that change in an impedance measurement, you may be able to tell something about the cells. The way people do tissue culture normally is to grow cells in an incubator for a few days, take them out and look at them. They photograph them and tell people what they see; whether see; whether they're big, small, or grow in a pattern. We wanted to find out if there is some way we can measure the condition of the cells without describing in general terms that happens. That is why we started out with this technique.

You have to design a new tissue culture dish. Figure 34 shows how it looks, with evaporated gold electrodes on the bottom, one big and four small ones. You have wires attached and put some red wax on to cover up your

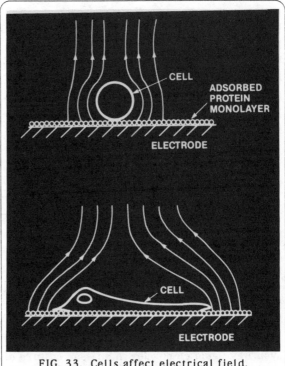

FIG. 33. Cells affect electrical field.

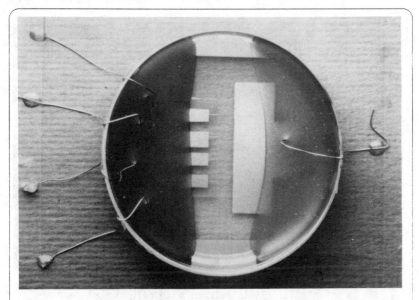

FIG. 34. Culture dish with electrodes.

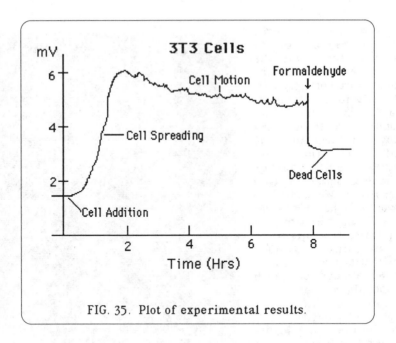

FIG. 35. Plot of experimental results.

solder joints because the cells are very sensitive to that sort of thing. Now you have gold electrodes in a tissue culture dish.

Let me show you the results of an experiment. It works very well. In Fig.

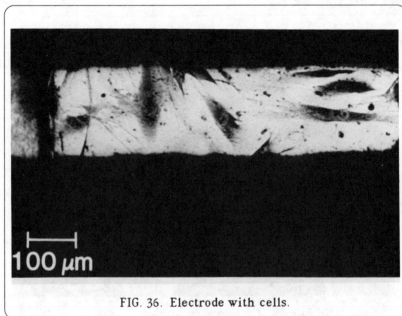

FIG. 36. Electrode with cells.

35 we plot time in hours along the x-axis, and millivolts along the ordinate (the in-phase voltage). As physicists you know that we can measure both in-phase voltage and out-of-phase voltage. You start out with maybe a few millivolts. As you add the cells, the cells drift down to the electrode and they start attaching. Resistance, or if you will, the milli-voltage at the electrode increases by a factor of 3-5. Then you get a little noise in the in-phase voltage. It is really not electrical noise, but it is because the cells are moving. Remember, the cells are alive.

Figure 36 shows an electrode with cells on it. This is the reason, the basic reason, why the resistance changes. When the cells attach to the gold electrode, they block some area. Therefore, it's natural that the resistance should increase. It turns out to be more complicated than that, but that's the first-order relationship.

As I told you, I was very much interested in the way cancer cells metastasize, which means that the cancer cells let go from where there are in the body and attaches some other place. Now, I can't really control the cells, but I can control the surface they grow on. We have started using various different kinds of protein. I show some data in Fig. 37. I already told you about BSA, but there are other proteins like fibronectin, gelatin, fetuin, and what not. There are hundreds of thousands of proteins really. We

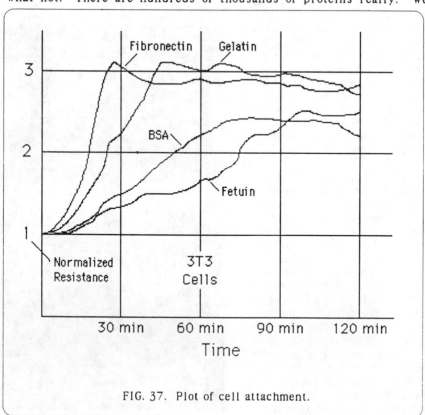

FIG. 37. Plot of cell attachment.

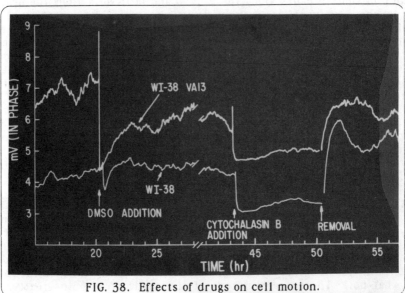

FIG. 38. Effects of drugs on cell motion.

found out that the cells like fibronectin very well; it attaches very fast and it doesn't like BSA anywhere near as well. You can compare normal cells with cancer cells this way, for example, finding out what they like to attach to. That's one direction this research has taken.

Another direction this research is taking is to look at the effect of drugs on the cells. Now we can do this in real time. In Fig. 38 are two traces of in-phase voltage. The upper trace is from a cancer cell and the lower is from it's normal counterpart, a normal cell. You get wiggles because the cells are moving. Then we apply DMSO, which is dimetyl/sulfoxide, which is the thing people rub on their tennis elbow. I wouldn't recommend that but that's one thing people use DMSO for, and some people believe it's a wonder drug. What you see is

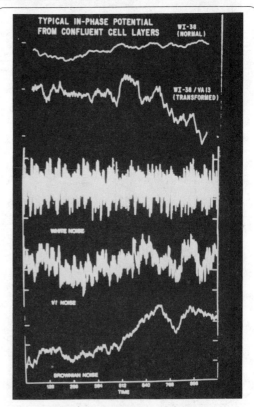

FIG. 39. Examples of noise.

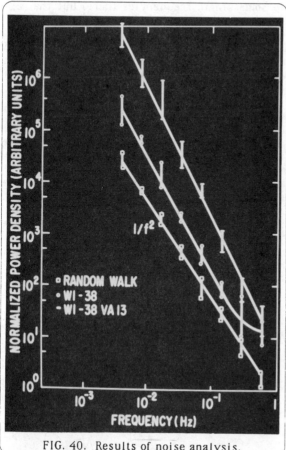

FIG. 40. Results of noise analysis.

that the DMSO will go into the membranes very fast. That has a very large effect on the cancer cell and a very small effect on the normal cell. You can measure that in real time. Then the cell will recover as you see from the graph.

Then, at some later time, we put cytocolasin-B, which is a drug that interferes with the muscle fibers in the cells. Therefore the cell motion charges very rapidly and the cells barely move. If you remove the cytocolasin B, the cell will recover and they go on their merry way. So anything you do from the outside, like changing the pH, changing temperature, changing medium, changing drugs and other things, you can measure in real time what happens with the cell. We are involved in that kind of work.

Another aspect I became involved with, but wished I hadn't, is the noise. If you look at these curves, your eyes see all sorts of cycles and periods. I thought it would be a wonderful thing to find all these periods, so I started analyzing the noise. That's almost the last stage of a physicist, when he analyzes noise. The last stage is when he becomes a historian. So I'm on my second to last stage by analyzing noise.

At the top of Fig. 39 is noise from normal cells, next is the noise from cancer cells, next is white noise, ordinary white noise. Next is 1/f noise. On the bottom is Brownian noise that goes as $1/f^2$. The noise from the cells look more or less like Brownian noise. Then you can make digital filters and analyze it. It takes a lot of time.

What you get out is unfortunately not very exciting. Random noise, simulated on a computer, gives $1/f^2$ noise. If I analyze the data from the cells, I get a little steeper slope from the normal cells and steeper still from the cancer cells, as indicated in Fig. 40. But you see they're not very drastic. There are no specific frequencies that come up and that's what I'm looking for. I can't find any and I'm very sorry about that. If I had, this would become very important, but so far I haven't. So now we are looking at single cells and hope we can pick up some single frequencies there, but that experiment is more difficult to do.

Actually, for those that are interested and for those of you who enjoyed the beautiful lecture by Professor Libchaber, this noise of course is fractal. That's a saving grace. If you have no frequencies in there you can always give it a fractal number and that's good.

Let me wind up on the talk with a few comments about "What is life?" Schroedinger wrote a very influential book where he talked about life. The book is somewhat ambiguous, but I think what Schroedinger said is that life can be explained by physics and chemistry. I will go one further -- I'm not quite sure what Schroedinger said, I read the book, but I'm still not quite sure -- but, I go one further and say that life really can be explained by the present knowledge of physics and chemistry. You don't need to know any more physics and chemistry. Actually, life can probably be explained by classical physics and chemistry. Arthur Schawlow said yesterday that people didn't take the Balmer series seriously. I think physicists don't take life so seriously, because it's too complicated. Then you ask, "If it's so simple, why don't you explain it to me?"

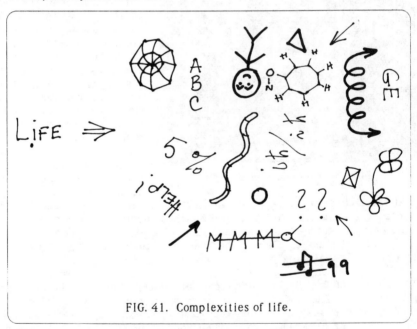

FIG. 41. Complexities of life.

Figure 41 is meant to show you that life is complicated after all. It is very complex and that's where the problem is. The way I explain it to General Electric managers is that life is like the stock market. Everybody understands how to buy bread, how to put money in the bank, how to spend more money than you make, and all this sort of thing. Everybody knows that, but nobody can put the whole together and predict what the stock market is going to do. Life is similar. We understand a large amount of detail about life, but it is too complex and people can't, at the present time, put it altogether into a whole thing. I don't think in my lifetime it will happen, but if you work very hard, it certainly will happen in your lifetime. Thank you.

This work was partially supported by the National Foundation for Cancer Research. This talk was based on an earlier talk at Philip Morris and published in Natural Products Research: The Impact of Recent Advances. The Proceedings of the Fifth Philip Morris Science Symposium. Philip Morris has given us permission to reprint the article.

MILDRED S. DRESSELHAUS:

I'd like to thank Dr. Giaever for his wonderful introduction to biology. He points out, as we will see several more times before this celebration is over, evidence about the influence of one field of science on another. It is very important to know what physicists have done for other fields of science and also what people from other walks of life have done for physics.

Our second speaker today is Professor Phillip W. Anderson, the Joseph Henry Professor of Physics at Princeton University. Phil is another renaissance man that seems to know more about any other field than many pros working in these fields. He always seems to be knowledgeable about anything I've ever asked him about. Phil is a person that has all his degrees from Harvard, never went any place else to study it seems. That's not really true. After completing his Ph.D. at Harvard, he moved to the Bell Telephone Laboratory, where he had a very illustrious career working in almost every branch of theoretical condensed-matter physics.

I remember myself as a very young graduate student reading some of his papers on magnetism and was convinced that that was an exciting field to work in. His contributions in condensed-matter physics, as I said, covered many fields: magnetism in the early days, disorder, the area for which, probably more than any, he got the Nobel Prize in physics in 1977. His more recent work, and even long-time work, covers the field of localization, superconductivity and, as I said, every branch of condensed-matter physics. For the last 20 years he's been a professor, setting up young physicists in theoretical fields around the world. First he was working at the University of Cambridge and more recently at Princeton. In the last he's gotten very excited by high-T_c super conductivity, with the announcement of Bednorz and Muller of this extraordinary event of superconductivity above $40°$ and $90°$, and then we'll find it's even higher than that. I don't want to steal the thunder of their talks, so let me, without further ado, introduce Dr. Anderson, who will give us a talk on "Strange Insulators, Strange Metals, Strange Superconductors: High T_c as a Case History in Condensed-Matter Physics."

Strange Insulators, Strange Semiconductors, Strange Metals: High T_c as a Case History in Condensed-Matter Physics

Philip W. Anderson

DEPARTMENT OF PHYSICS, PRINCETON UNIVERSITY

Princeton, New Jersey 08544

Actually at the time that high-T_c superconductors happened, I was hoping to escape from condensed-matter physics into some of the implications of spin glasses and neural networks and various other things. So it was really with considerable regret at first that I found myself driven back into my old stamping grounds.

The Michelson-Morley experiment is one of the best examples, of the importance of true simplicity in physics. It shows how a simple fact can have profound implications on the laws of physics, and how terribly simple those laws of physics are. That underlying simplicity leads to a tremendous amount of complexity, as Ivar Giaever was describing so ably in the last lecture, and leads even to the complexity of living things. Biology is an enormously complex and exciting field, but one in which, every once in a while, you can return back to simplicities. The key to ever understanding life with the human mind is to find some of the underlying simplicity as Ivar Giaever very clearly has done.

This duel between simplicity and complexity is the fascination of condensed-matter physics. You start from simplicity and the fundamental laws, applying these to complex systems, and then find that in the end some new simplicity comes forth and allows to you understand what is going on in what, at first, looks like a bewilderingly complicated system. I had meant, when I first accepted this invitation, to give you an overview of the many exciting things that have happened in the past decade or so in condensed - matter physics, most of which have that kind of character, but then high-T_c intervened. This phenomenon, in itself, contained so many bits of condensed-matter physics, that I felt that it was, in itself, a perfectly adequate illustration in a general sense, of how the field works.

First of all, I just want to make some acknowledgments because I don't want to pretend that I'm working alone, or that anything I'm talking about is the result of a single person's work. There's a lot of groundwork by people like Ramakrishnan, Affleck, Rice, a lot of work simultaneously that has been helpful by Kivelson, Rokshar, and Sethna and by Langhlam. There's almost too much generous sharing of experimental data, by now maybe 10 or 20 preprints a day of it. In fact, in the beginning of this work some specific friends from the experimental world gave me some very important data and recently I've been depending very heavily on data from a number of other people so I mention these people specially here.* Most of all, there's a wonderful group of students and junior people at Princeton with whom I've been collaborating right through the whole enterprise.**

Now, I want to emphasize two things about this problem. Both of them center around the fact that before you can solve the obvious problem, you have to understand what is underneath the obvious problem. You have to understand the substrate of what you're doing. Let us start from a cliche (Fig. 1) about what is going on in these materials. Two years ago this would have been considered to be not just a cliche, but a truism. Unfortunately, it's not true.

The questions one should ask are: what forces, are there any such forces; what electrons are there, what is the normal metal, and is there an energy gap? As far as I'm concerned, the statement in Fig.1 has something like

> SUPERCONDUCTIVITY
> IS DUE TO ATTRACTIVE
> FORCES WHICH PAIR
> ELECTRONS IN THE NORMAL
> METAL LEADING TO AN
> ENERGY GAP.
>
> **THIS IS FALSE.**

FIG. 1. Superconductivity cliche.

four operative words, of which three and a half are wrong. A more or less correct statement, is probably something like that given in Fig.2.

> SUPERCONDUCTIVITY IS
> DUE TO INTERLAYER
> TUNNELLING, WHICH
> PAIRS HOLE SOLITONS IN
> THE "NORMAL" METAL
> LEADING TO AN ENERGY
> GAP
> (for some of the particles).

FIG. 2. " Correct" representation of superconductors.

Superconductivity in the end is due to a rather unexpected part of the problem, namely the tunneling between the layers of copper atoms. The carriers of charge in the normal metal are not electrons, they are some kind of soliton. The pairing is undoubtedly there and leads to an energy gap perhaps, but certainly only for some of the particles.

To understand the background of these rather unexpected statements, I have to point out that underlying the phenomenon of high-T_c superconductivity, which in itself is a very fascinating thing, there are two stages of understanding of just what the normal materials are about (Fig. 3).

* N.E. Phillips, L. Greene, R.C. Dynes, N.P. Ong, Paul Grant, D. Bishop, G.A. Thomas, among others.

** G. Basharan, Z. Zou, T. Hsu, J. Wheatley, B. Boucot, S-D. Liang, R. Kan, T. Wen, E. Abrahams, S. Coppersmith, B.S.Shastru, S. John.

Namely, we have to understand a very strange insulator, which hangs around in the background of these materials, and then we have to understand a very strange metal from which the superconductors arise (Fig. 4).

There are first mysterious insulators. The physics is best illustrated by the very first of the superconductors, which is much less complex, namely the lanthanum - copper - oxide - based one. Right from the start, one saw that lanthanum-copper oxide itself had a number of mysterious properties even though it was not only not superconducting, it was not even conductive of electricity.

There are various ways of making a corresponding insulator in the yttrium-barium materials, the other high-T_c materials; I have quoted only one of them. Then one dopes these materials, one changes them by adding some percentage of carriers and arrives at the famous material ($YBa_2Cu_3O_7$) which is often called, and which I will probably always call, *123*.

These are very mysterious metals when you look at them. They have very strange resistivity properties; their susceptibility is not exactly what you would have expected either.

---

(1) STRANGE INSULATOR: QUANTUM SPIN LIQUIDS=RVB.

(2) THE NORMAL METAL IS NOT "NORMAL."

FIG. 3. Two stages of understanding of normal materials.

MYSTERIOUS INSULATORS: La_2CuO_4
$Y(La_2)Cu_3O_{6.5}$

MYSTERIOUS METALS: $(LaSr)_2CuO_4$
$YBa_2Cu_3O_7$

MYSTERIOUS SUPERCONDUCTORS.

START FROM BEGINNING.

WHY INSULATORS?

WHAT KIND OF METAL? WHAT ELECTRONS?   WHAT FORCE?

IN THE BEGINNING IS A MYSTERIOUS INSULATOR WHICH SPAWNS SOLITONS.

SOME DICHOTOMIES WHICH MUST BE PASSED BY (MOSTLY) CHEMICAL REASONING.

FIG. 4. Mysteries and questions.

Figure 5 shows you that there is a really mysterious phenomenon going on here, although I'm showing you this in a way which is mostly comprehensible only for solid-state physicists.

Figure 5 depicts the resistance behavior of one of these materials. For comparison, I show not an ordinary metal, but Nb_3Sn which has a very peculiar behavior of the resistivity. The main thing to note is that the resistivity of Nb_3Sn is considerably less than that of the high-T_c superconductors. At low temperatures Nb_3Sn has quite a bit of curvature, at high temperatures it saturates. Other metals on the same scale would look much

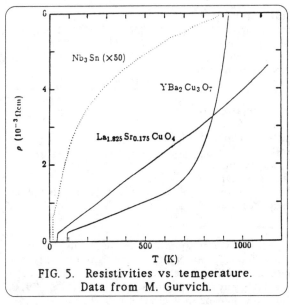

FIG. 5. Resistivities vs. temperature.
Data from M. Gurvich.

more like an upward curvature at first, then a linear behavior, but they would have much more detail in them and the magnitude would be totally different.

By contrast, the lanthanum - strontium - copper material is, as we can see from Fig. 5, linear in temperature up to about 800°. The resistivity rises, and it rises in fact not because the intrinsic behavior changes, but because you can't keep the oxygen in the compound above that temperature. The same thing is true of the 123 material, it is linear up to 600°, and then starts to rise only because you can't measure it. I think I have an idea from other measurements that this behavior continues least for another factor of two in temperature.

These metals are not your usual metal and one really shouldn't begin thinking of them as though they were, until one has solved the problem that one doesn't understand why lanthanum-copper oxide is an insulator, and doesn't understand its state. Then when you dope it, you don't really expect, until you understand why it was an insulator, why its resistivity is very strange.

So we have this problem of why are they insulators and, then, what kind of metal? And so, in the beginning is the famous mysterious insulator which, I will show you, spawns the concept that the system has to have solitons (Fig.4).

Before we get to the mysterious insulator, we have to have a little understanding of what all of this phenomenology is happening in. To do that you mostly have to use common sense, which is often in physics known as chemical reasoning (Fig. 4), and therefore looked down upon. It is always a mistake to look down upon anyone, and particularly when studying complex materials its a mistake to look down upon chemists and upon materials physicists, who have thought hard about what the physics of these oxide materials is. What you will notice about these materials when you look at them right off the bat (I'm not going to show a picture of the the crystal structure because I'm sure Paul Chu will do that later and, in any case, its very complicated.) is that all the superconductors contain a set of copper-oxygen planes that are very rigid in a certain sense, very rigid in that they do not like at all to have the copper-oxygen distance vary by one iota. They do all kinds of wiggles and waggles and twitches, and so on, transformations that change, perhaps, the geometric structure of these layers, but the layers remain absolutely rigidly square-planar in their topology.

These square-planar layers have Cu^{++} ions in them, and Cu^{++} ions contain 9 d electrons and there's no doubt about that. (d)^9 is the valence of copper in copper sulphate; we're very familiar with how divalent copper behaves. A (d)^9 ion has an odd number of electrons, and you say, well, if it's got an odd number of electrons it has to be a metal because there's no way I can make an insulator obeying band theory. So, right off the bat you say, this is a so-called Mott magnetic insulator, namely, the electrons are kept apart by very strong repulsive forces. So the first dichotomy you will arrive at, not really by chemical reasoning but by primitive common sense, is to say this isn't a metal, the insulator, at least, is a Mott magnetic insulator.

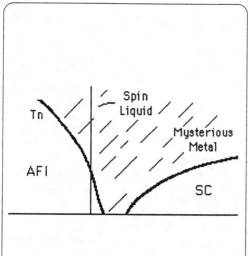

FIG. 6. Generalized phase diagram with insulator antiferromagnetic (AFI) and superconducting (SC) phases.

Then, what kind of a magnetic insulator is it? Is the important insulator simply a typical antiferromagnet? You go through the old Slater argument, saying the spins are alternating and that introduces an extra degree of periodicity and reduces the periodicity of the lattice and thereby you have bands which can have a gap. The answer is yes, there is an antiferromagnet in the phase diagram, but if one looks at the generalized phase diagram of all these materials in Fig. 6 (I assure you my sketch in Fig. 6 is a pretty good description of the phase diagrams that one actually has.) there is always an antiferromagnetic insulator on the undoped or negatively doped side - for instance, YBa_2CuO_6 .

It occurs at quite low temperatures and is very sensitive to composition. If you increase the doping, increase the metallicness, the phase which becomes the mysterious metal is the one that is above that phase transition, or the one which extends down to absolute zero slightly to the right of it.

So the superconductor grows, not out of an antiferromagnetic insulator, but out of a different insulator which, I'm going to argue later, is a certain kind of quantum spin-liquid. So the antiferromagnet is there but, by looking at the phase diagram in detail, you see that it has nothing to do with the interesting properties. It does of course have something to do with them, it tells you that it is a Mott magnetic insulator which, in turn, tells you that the force is dominantly repulsive, that you don't have to look for a strong attraction because it can't be there.

Finally, the question is what bands are these mysterious carriers in? There you have to use two bits of chemical reasoning. One is the retention of square planar shape and the other is the concept of Wannier functions, which chemists call symmetry orbitals. The concept of symmetry orbitals-- its a little primitive arithmetic, you have to be able to count up to ten in studying d electrons--a little primitive arithmetic tells you there is only one band that they can be in. They can be in a certain symmetry orbital centered around the copper lattice. When you come through with those three fundamental and simple bits of reasoning, you conclude that there is only one model that you should be thinking about, the so-called Hubbard model. In order to understand the peculiar insulator, you should be looking at the Hubbard model for a half-filled band.

I've drawn a primitive kind of picture here for the Hubbard model, namely in these square-planar lattices I've got a square-planar group of coppers.

I've removed the oxygens, not because they aren't there, but because the symmetry orbital is centered on the coppers. I could make it centered on the oxygens if I had to, but it would just make my thinking more complicated. Many people make a fuss about the conducting holes being on the oxygens, but the vital question is really their symmetry, not which atom they are on.

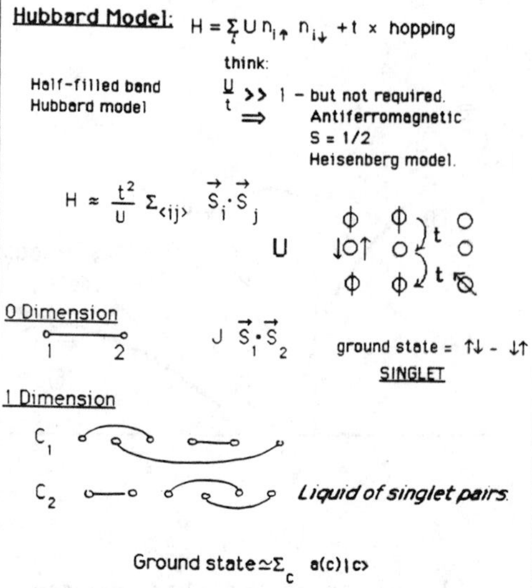

Hubbard Model: $H = \sum U\, n_{i\uparrow}\, n_{i\downarrow} + t \times$ hopping

think:

Half-filled band
Hubbard model

$\dfrac{U}{t} \gg 1$ - but not required. $\Longrightarrow$ Antiferromagnetic $S = 1/2$ Heisenberg model.

$$H \approx \frac{t^2}{U} \sum_{\langle ij \rangle} \vec{S}_i \cdot \vec{S}_j$$

ground state $= \uparrow\downarrow - \downarrow\uparrow$ **SINGLET**

$J\, \vec{S}_1 \cdot \vec{S}_2$

0 Dimension
1   2

1 Dimension

C_1

C_2 *Liquid of singlet pairs*

Ground state $\simeq \sum_c a(c) |c\rangle$

FIG. 7. The Hubbard model.

The Hubbard model (Fig. 7) is a model in which there is a hopping integral from site to site, and there is a strong repulsion, U, whenever two electrons sit on the same site. If you think hard about a system where electrons can hop from site to site, but there is a very strong repulsion preventing two electrons from ever sitting on the same site, and you have a number of electrons exactly right for the count of sites, its not going to be able to conduct electricity at all because absolutely every site would be occupied with exactly one electron, which has a spin. The effective Hamiltonian, it turns out, was derived way back in '59. It involves an antiferromagnetic interaction between the spins on the individual sites, and nothing else.

There is a lot of old theory about antiferromagnetic systems. The question of whether antiferromagnetic systems have an ordered ground state has a history so venerable you can hardly believe it. The first person to approach it was Hans Bethe; the next person L. Hulthen, then H. A. Kramers,

the next person -- well, after that, I guess I was maybe one of the next people, except for Kubo. Actually it was a very important moment to me when
I learned H. A. Kramers had also been working on it; he thought my work
was very useful and invited me to come discuss it with him, when I was, really, just about so high, at least so far as physics was concerned, and I was
very proud of that. Kramers was very interested in this problem, he knew
it was a problem.

The reason it is a problem is that if you start from simple systems, like
two spins, or one-dimensional chains, you get one type of answer. In "ze-
ro-dimension," i.e. two sites, you know what happens, you get a singlet
ground-state of the two spins (Fig. 7). The singlet ground-state has a re-
markably low energy: It is very energetically favorable because all three
of the components of the spins manage to line up antiparallel giving an
energy $(\sigma_i \cdot \sigma_j) \to -3$, whereas if
you only line up one of the com-
ponents antiparallel you only get
a factor -1. So you gain a factor
$(1+1/S)$. This factor fundamentally changes the ground state of the
one-dimensional system: that
ground state was derived by Bethe,
and commented upon by many
other people, up to the present,
e.g., the work of Fadeev, Affleck,
Holdane, etc.

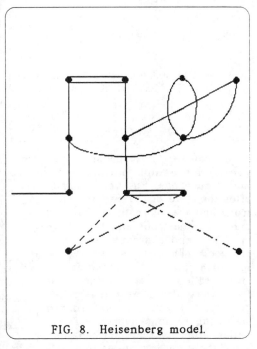

That ground state is pretty definitely a liquid of singlet pairs, a
liquid in which singlet pairs move
around freely, interchange freely, from one pair of sites to another (Fig. 7). So the question is, does
there exist the same corresponding resonant motion of valence
bonds within the two-dimensional
system. That is the thing that I
call a resonating-valence bond
state, but I'm perfectly happy
with any name for a state which is

FIG. 8. Heisenberg model.

homogeneous, a singlet, and has no broken symmetry, and that it is the
ground state, or nearly the ground state, of the two-dimensional Heisenberg model (Fig. 8).

It certainly is the state at finite temperatures because the two-dimensional Heisenberg model cannot order at finite temperatures. Pictorially,
the state that I want to describe is the sum of all possible singlet configura-
tions of bonds on the square lattice. I have drawn a number of these in
Fig. 8.

Incidentally, I should add, as you go to higher dimensions or higher
spins, this lovely thing disappears and you get a boring state, the boring

In the Beginning:

God made the electrons

$s = \frac{1}{2}$ and He looked, and said they were good.

(-e)

Then Bob Schrieffer made the solitons

$s = \frac{1}{2}$ (-e)

and P.W.A. looked, and said they were better; BUT THEY ARE STUCK IN PLANES.

FIG. 9. Genesis.

state being a state in which the spin on one such lattice is up, the spin on the next is down, up, down, up, down, et cetera, i.e, in which the spins merely alternate. That's called a *Neel state*, and is valid probably for most systems in three dimensions, but not all systems, and for high enough spin in two dimensions in the ground state. But I'm almost certain at this point it is not the ground state of the two-dimensional square lattice with S-1/2. That statement is new and depends on a number of theories, all of them unpublished, by different people around the country -- the world, sorry (one's in Canada).

Well, what are the excitations of this beautiful state? It turns out that it has two kinds and that they are very interesting.

Of course, electrons are interesting (Fig.9). Michelson and Morley, I guess, didn't think much about electrons, in fact, probably didn't know about electrons, but very shortly after them we discovered the electrons and we all realized that God made the electrons and they have spin 1/2 and charge e, and they are good. Then it was discovered in some dirty, but fascinating stuff, called polyacetylene, that there were also some things called *solitons*. If you have an ordered state, the solitons have the fascinating property that you can separate the spin and put it in a different place from the charge. You can have particles, one of which has the spin and the other which has the charge. Bob Schrieffer, among others, first made these. I think for high T_c superconductors they're much better than simple electrons.

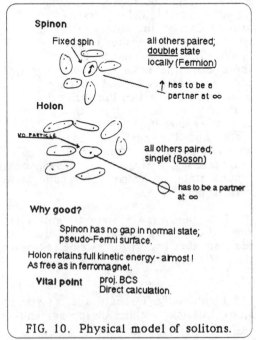

Spinon

Fixed spin   all others paired; doublet state locally (Fermion)

↑ has to be a partner at ∞

Holon

NO PARTICLE

all others paired; singlet (Boson)

has to be a partner at ∞

Why good?

Spinon has no gap in normal state; pseudo-Fermi surface.

Holon retains full kinetic energy - almost ! As free as in ferromagnet.

Vital point   proj. BCS Direct calculation.

FIG. 10. Physical model of solitons.

I'm just going to give you a pictorial model. We have lots of theory, but it certainly is not suitable for an audience this size.

The first and simpler soliton, the one that came out of our original theory, is what we call a *spinon* (Fig.10). Which is this: You take this mess of singlet pairs and you spread it away, you push it away. That pushing it away is a nontrivial process, because its incompressible, and you're going to push part of it all the way to infinity. So, that part of it as you push it away leaves an extra partner particle, one extra electron appears at infinity, but anything at infinity can obviously be ignored so we are going to ignore that.

Then on the site that we cleared off, we are going to allow ourselves to have a single fixed spin. That particle looked at locally has Fermi statistics, but it obviously has no charge because I haven't emptied any sites. So its a fermion and locally its a doublet state. I can imagine bringing in or removing that electron now. Just take away that electron, still leaving the partner at infinity, and leave no particle there. Since that involves removing a fermion, that means that that is a singlet particle, its a boson, and it has a charge of e^+. Now, if the ground state was this, and if these particles are there, why are they useful?

The answers to these questions are very important. One of them is that the spinon state has no gap in the normal state. These funny fermi particles certainly have no gap and probably even have a fermi surface, a whole surface of states in k-space. Whereas the *holon* (Fig. 10), if you build it correctly, has much lower energy than you would have if you put a single electron in there because it turns out that you can build a holon--build a boson, not a fermion-- build a holon, in fact a whole gas of holons, that retains just as much kinetic energy as though it were a free particle.

Well, we arrive at these by at least three independent approaches to the mysterious insulator, again I'm not going to bother you too much with theory. These three independent approaches all arrive at the same result, but much more important, of course, for any physicist, not necessarily for a mathematician, is that there are two experiments which agree with this prediction of a pseudo-fermi surface. One is the neutron inelastic-scattering experiment of Endoh, Birgeneau, *et al.*, which was recently published in *Phys. Rev. Letters* **59**, 1613 (1987). Another is the fact that everyone who has so far measured the insulators' specific heat at low temperatures has found that it looks like a metal as far as specific heat is concerned. It has a linear specific heat at low temperatures which is so big that it is very hard to explain in any other way than just a pseudo-fermi surface.

Now we have the carriers and can get the mysterious metal out of this . I showed you the data (Fig. 5) already of the very strange resistivity properties. The theory tells us that the low-energy carriers which compensate the charge are these bosons (Fig. 11). We can think of them as bound states of a charge and a spin. They are bound by the kinetic energy. You can think of them, essentially, as what you have to do in order to put a particle in the top of what is called the lower- Hubbard bond, in order to regain the kinetic energy you lose if you try to put a pure electron in. This odd pic-

ture, this very peculiar picture, is
verified by all the transport results that we've been able to find.

Shown in Fig. 12 are the results
of the first single-crystal resistivity measurements. It was done by
the IBM group and this already is
very strange. Figure 12 provides
my second exhibit for the proposition that this is a very mysterious
metal. We're measuring the resistivity on the left-hand scale parallel to the planes of a single crystal, and on the right- hand scale
perpendicular to the copper-oxygen planes. The resistivity parallel to the planes does look very
linear, it has a small intercept, but
mostly it's linear. As you saw in
an earlier figure it goes on being
linear up into the sky. In fact, if
you took one- quarter of a percent
of the c-axis values and added (you
cannot place your electrodes that

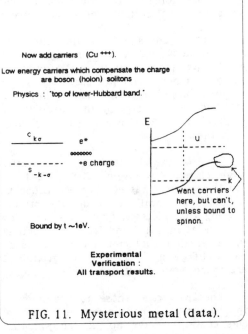

FIG. 11. Mysterious metal (data).

accurately), you would find that it changed to having absolutely no intercept.

On the other hand, the resistivity in the "c" direction is a hundred times
as high (I will show you better data, more modern data, in which it is a

thousand times as high)
and which has the oppo-
site temperature depen-
dence. Namely, it be-
haves more like a semi-
conductor than it does
like a metal.

There is a fascinating
way in which you can
straighten out the ex-
perimental data on the
resistivity and make it
look much more ratio-
nal. This is done in Fig.
13. One takes those re-
sistivities (this isn't
those resistivities, but
some that Phuan Ong did
on some of the better
single crystals), you
find that if you plot the

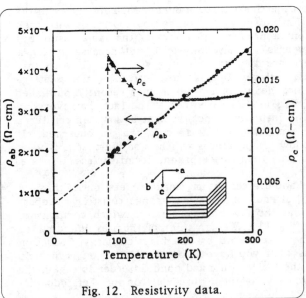

Fig. 12. Resistivity data.

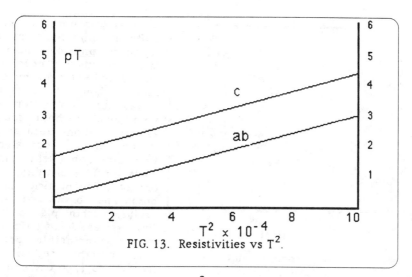

FIG. 13. Resistivities vs T^2.

resistivity times T as a function of T^2, you get an absolutely straight line.
(Figure 5 refers to the c-direction, but actually one of these electrodes was
set in the ab-plane.) You get, in fact, an equally good straight line if you
plot either of these curves. If you think about that, that says that the resistivity in the c-direction is proportional to $1/T$ -- no curvature, no nothing -
- and the resistivity in the ab-plane is directly proportional to T.

Incidentally, just so that you can see that these are real genuine materials that we're talking about, Phuan Ong has given me one pretty picture,
shown in Fig. 14, of the growth of a single crystal in a copper-oxide base
flux. Figure 14 shows a single
crystal of this 123 material. They
are nowadays quite easy to grow.
In even the Princeton physics
lab you can make one-millimeter
crystals and in really good places
that know how to do these things,
you can make one-centimeter
crystals and do some real physical
measurements. Ong, in fact, was
the one who provided me with the
nice data in Fig. 15.

There is just no question but
what the resistivity is proportional to T in the ab-directions
and it's proportional to $1/T$ in the
other direction (Figs. 13, 15). The
theory that gives that is a very

FIG. 14. Picture of 123 material.

straightforward affair, in which you ask what would happen if I had a fermion scattering against a boson, the boson being the charged particle (Fig.
16).

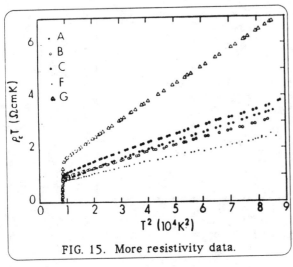

FIG. 15. More resistivity data.

This was done, in fact, quite a while back, by a Japanese physicist named Maekawa and it's been repeated and extended by my student Zou. The answer comes out off the top of your head without doing any calculations because it is well-known that electron-electron scattering leads to a T^2 resistivity, but one of these particles is a boson and doesn't obey the exclusion principle so the final states don't have an exclusion-principle problem, and that reduces it from T^2 to T, so one gets a linear-in-T resistivity.

Incidentally, because the matrix element which causes this scattering is quite large in our calculations, the resistivity by this spinon process is going to dominate any normal phonon resistivity, you're not going to see anything else. Everything else should be about an order of magnitude lower and should be in the ballpark of that niobium-tin resistivity, which as you can see (Fig. 5) is fifty times smaller, so you wouldn't see it. The coefficient you can work out, and is about right for what you might expect.

On the other hand, if you look at the resistivity in the other direction, you recognize a fascinating fact. If the current is carried by these particles, the funny particles, these beautiful solitons that have the charge separated from the spin, it can't get from one layer to another, it's stuck. The solitons are excitations of a single layer, they can't live without their partner at infinity; in some sense, they can't live alone and they certainly can't go out into empty space (Fig. 10). Flux lines live only in superconductors, and, correspondingly there are no solitons in empty space. In particular, in the empty space between the copper-oxygen layers there are no solitons, you have to

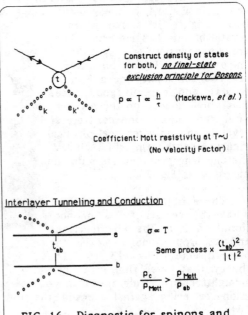

FIG. 16. Diagnostic for spinons and holons.

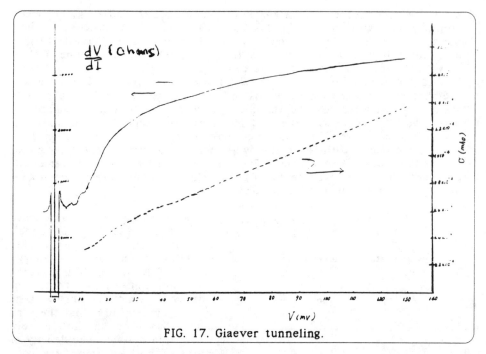

FIG. 17. Giaever tunneling.

put an electron together in order to get back and forth. So you have to combine a boson and a fermion, make them into an electron, then let them tunnel and come into the next layer.

The effect is, that except for changing resistivity to conductivity, you get exactly the same calculation. Its basically a scattering calculation of boson against fermion, but it has an extra factor of the tunneling matrix elements squared between the layers. When you look at the numbers, in the first place, you check the basic inequalities they must satisfy, but also, it tells you that the tunneling matrix element must be something like one one hundredth, that is about a hundred times smaller hopping integral between the planes. This is a perfectly reasonable hopping integral given the geometry.

This would be to me very convincing, but I realize you may like something a bit more explicit. This isn't even necessarily more explicit, but it's fun, because it's tunneling. Figure 17 depicts Giaever tunneling between a lead slab, or a lead film, and the 123 material, as measured by Bob Dynes. He feels that this must be a very perfect junction because he sees the lead gap down here, and he sees all the lead phonon spectra. So that this is a perfect tunnel junction as far as the lead is concerned. This measurement is made at $4\,^{\circ}K$, where the lead is superconducting.

The unusual thing is the background conductance, dT/dr, is linear in $|V|$. This is absolutely universal to all of the tunneling measurements that have so far been made. Nobody, as far as I know, except maybe at most one or two experiments and those not reproducible, has ever been able to establish

what the tunnelers call a background. They always find that the background varies with voltage in this interesting way. Actually to see from
Bob's curve in Fig. 17 how it varies with voltage you simply have to turn it
over and call it a conductivity and it is absolutely a straight line with a
small intercept. I thought the intercept was an artifact until we calculated
it, and that checks out too, from the fundamental theory. I have made
sketches of the many, many kinds of tunneling curves that people have
seen. The beautiful BCS-like tunneling curve is from a "typical" sample (I
asked the IBM people what they did they mean by a typical sample and they
said one in a hundred.). Anyhow, you look at various other data, other tunneling data, along with the scanning tunneling microscope, and you obtain
results that looks like the data in Fig. 18. Everyone has a lot of linear background, with maybe a few features in the middle. IBM actually sees a gap,
but above the gap, lovely linear conductance.

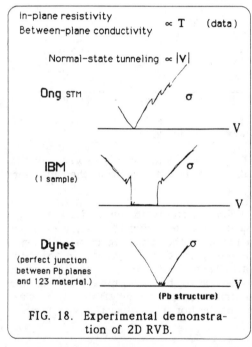

FIG. 18. Experimental demonstra-
tion of 2D RVB.

So, what we have to assume is
that the linear term is the tunneling into the normal metal, i.e.
tunneling from the high-energy
region of lead, which is normal,
into the high-energy region of
this stuff, which is normal (Fig.
19). Again, we infer that the tunneling has to occur into a hole
plus a spinon state. The density of
states of both of those are constant, so the current is the product
of the two densities of states
which is V^2; the conductivity, the
density of states, is V.

Since one knows about what the
tunneling experiment measures,
that it measures the imaginary
part of the Green's function, G_1,
which is basically linear in frequency, that tells you there is
something that is not an electron,
that the electron density of states
vanishes essentially at the fermi
level. There are no electrons there. This is the unequivocal measurement
that says there are no electrons at the fermi surface.

So, what does happen? That's being worked out. I like to describe our
situation as being at the Cooper-pair level (Fig. 20). What we're trying to do
is to see how the pairing comes about. There are two ways that you can
tunnel through the insulating layer between the metal layers. You can put
a single particle and its spinon back together. That we've seen. That's boring. (It's not boring, its interesting, but it gives you that linear V and it will
never lead to superconductivity.) The other thing is you can make a pair of
holes, then one of these singlet-pair holes can jump as a unit. Pair tunneling doesn't have any of these problems with solitons. Two solitons are two

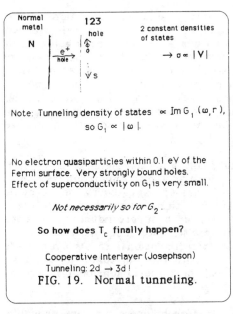

Figure 19 (left panel):

Normal metal

123

hole

2 constant densities of states

N

e^+ / hole

$\to \sigma \propto |V|$

$\sqrt{s}$

Note: Tunneling density of states $\propto$ Im $G_1(\omega, r)$, so $G_1 \propto |\omega|$.

No electron quasiparticles within 0.1 eV of the Fermi surface. Very strongly bound holes. Effect of superconductivity on G_1 is very small.

Not necessarily so for G_2.

So how does T_c finally happen?

Cooperative Interlayer (Josephson) Tunneling: 2d $\to$ 3d !

FIG. 19. Normal tunneling.

Figure 20 (right panel):

a = SC : Δ
b = ?
c = SC : Δ

Assume pair amplitude $\Delta = \langle e^*_k e^*_{-k} \rangle$ in layers a and c, calculate in b, make self-consistent.

Singlet pair
Hole pair

Tunneling Energy = $t^2_{ab} \Sigma \Delta_{ij}$

$$T_c \cong \delta \cdot |t_{ab}|^2 \Lambda$$

(Repulsive forces lower, but can't eliminate.)

$$E_k^2 = (\epsilon_k - \mu)^2 - (\Delta)^2$$

FIG. 20. Theory of Tc at the Cooper-pair level.

solitons are two solitons. You've brought the partner back, or the partners at infinity can cancel, and the two solitons can live in the intervening space so you can have Josephson-pair tunneling in the RVB state, whereas you can't have normal tunneling; normal tunneling is inevitably an inelastic process. So, you take Josephson-pair tunneling and you say, supposing I had two layers a and c that are superconducting and have pair amplitudes, let's calculate self-consistently the pair amplitude in a center layer b and then you try to make that self-consistent.

The trouble is we know very little really about the spectrum of these bosons, except that at high energies it is what you expect, k^2. We suspect at low energies it flattens out. We don't know exactly what kind of spectrum to put in. My best guess for what we put in comes out that the transition temperature is basically proportional to the pair-tunneling amplitude itself. T_c then becomes a number like t_{ab}, which is in the right ballpark (Fig, 20). If you use a roton spectrum you get a square root of tunneling instead. One doesn't really know. Given our present lack of knowledge of the precise values of these coefficients, both possibilities those fit the observed T_c's and both of those can be made quite large.

The other funny thing about this theory is that the boson quasi- particle energy doesn't behave the way a fermi quasi-particle energy does with a BCS gap equation. Because it's a boson you get a funny minus sign in there. The quasi-particle energy is given by E_k in Fig.20.

At T=0 the chemical potential can become equal to Δ and at T=0, then, you may have no gap for the charged excitations, but as far as we can see you would then have a gap for the spin excitations. It is also possible to arrange that $\Delta > \mu$ in which case there is a charge gap, but no spin gap

I emphasize that at this point everything depends on calculations of the exact low-energy spectrum of the bosons which we haven't really settled down upon. At this point we know to an gnat's eyelash what's going on in the mysterious insulator. We have a pretty good idea what's going on in the mysterious metal. We have a hypothesis for the mysterious superconductor and, just to further compound confusion we have an even wilder hypothesis for the strange unreproducible observations that people keep seeing above the transition temperature of the normal material, starting with Paul Chu's own measurements which showed that there was something strange happening at 240 °K.

The question is, is this normal metal really a normal metal, or is it, since it's a two-dimensional layer of bosons, is it in fact a superconductor, but only a two-dimensional superconductor? Present day theory isn't capable of answering that question as yet. There is a thing known as a Kosterlitz - Thouless state, a type of topologically- ordered superconductor that does exist in other cases, and we can't exactly see why we shouldn't have a topologically-ordered superconductor in our two-dimensional sheets even up to quite high temperatures. We can see that until the pair tunneling can take place, we're not ever going to have real, or three-dimensional superconductivity.

If we had a batch of two-dimensional superconductors our vortices would be point objects in the individual planes and they would have nothing but magnetic forces coupling them between the different planes. These magnetic forces are unbelievably small, microdegrees, so the vortices are very free to move around from place to place. The only way you could actually

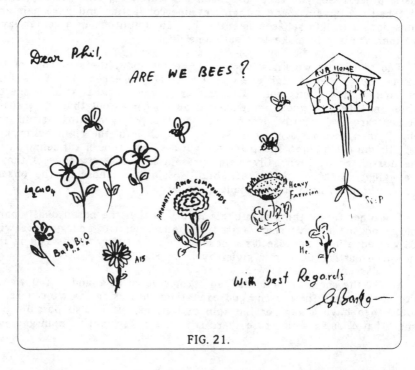

FIG. 21.

tell whether this is true is somehow to measure with nearly zero current and nearly zero magnetic field. It is perfectly possible, as the wildest speculation I can possibly leave you with here, that every sample of 123 is a room temperature superconductor.

I close with a postcard (Fig. 21.) that my collaborator, Baskaran, sent me from Aspen. He's Tamil and he doesn't pronounce the English language the same way the rest of us do, so RVB's becomes a question - and this is the answer. As far as we know, certainly these are B's, but maybe there are some B's hanging around in a lot of other interesting places.
□□

Grand Challenges to Computational Science

Kenneth G. Wilson

CORNELL CENTER FOR THEORY AND SIMULATION IN SCIENCE AND ENGINEERING

CORNELL UNIVERSITY

Ithaca, New York 14853

I'm going to violate the spirit of the symposium because this has been mostly devoted to the hundred years since the Michelson-Morley experiment. I certainly agree with everybody else in the symposium that the Michelson-Morley experiment was a great watershed in science. But in my talk, I'm going to go back four hundred years and I'm also going to go forward four hundred years. As I said, that is violating my instructions.

In the period from the time of Galileo, we have seen the development of experimental science, which was invented four hundred years ago, to the extraordinary situation we have today, where even after four hundred years we have revolutions taking place in astronomy, biology, in the whole area of atomic and molecular science, based in part on revolutionary advances that continue to take place in the microscope and the telescope, the epitomes of experimental instruments.

Both the microscope and the telescope were invented four hundred years ago. If you think of what it was like when Galileo was doing his first experiments, of course you can't see anything in the writings at that time that could have anticipated the richness of science as we see it today based on the current capabilities of microscopes and telescopes.

Furthermore, when you look at what was actually going on when experimental science got its birth in terms of the experiments being done by Galileo and his graduate students, the things that they were doing looked pretty stupid. If you don't agree with that statement, think back to your own experiences as freshmen in college or in high school physics doing those same experiments, and you certainly thought they were stupid. Furthermore, it is for certain that the establishment, the scientific establishment of that time, thought they were stupid, and wondered why funds were being diverted from the Medici Foundation to pay graduate students, who simply sat there in front of the tower of Pisa watching an hour glass.

Our perspective is different, of course, because we can look back on the developments that have taken place since the time of Galileo and recognize that is when it all started. Now, today we are looking at a third mode of science, computational science. It has the same problems that Galileo faced, but even worse, because now we have the richness of experimental science, of the discovery of the laws of the heavens and the observations in the heavens, the discovery of the existence of electrons and nuclei and all the richness of quantum mechanics. So to do the same kind of stupid things that Galileo was doing in 1590 is even harder to justify when you have to do it now, I tell you. When you watch a scientist interested in computational science writing a program in Fortran, it is indeed just as stupid as watching that hour glass in 1590. And we hear about it from our colleagues in exactly the same way that Galileo heard about it.

So what I want to do is to take a view of what's happening today and what will develop in the near future in computational science, but try to anticipate how this whole development will be looked at four hundred years from now. Now there are various uses of computers in science today. It's most important use is as kind of assistant, as in high-energy experimental physics. The high-energy experimental physicist would not be very happy if something happened to the electric services and none of the computers worked anymore. On the other hand, that is not what one talks about when one talks about a great experimental discovery in high energy physics. You talk about the W particle and you don't usually spend very much time talking about all the on-line and off-line computation which was necessary to identify the particle. I think as one looks back from the future, the real importance of large scale computation will be in computers as complete scientific instruments in themselves, rather than as servants to experimental science.

Let me say a little bit about what I call the supercomputer as a scientific instrument and let me say why I introduce the supercomputer in this context. Supercomputers are technologically the most powerful computers available for scientific and engineering calculations at any given time. There are certain companies that make what are called supercomputers today. I don't want to go into that because the reason I want to talk about supercomputers is that I want to anticipate the capabilities of computers tomorrow. What I'd really like to do is to anticipate the kind of computers that will be available four hundred years from now. But it is a little hard for anyone to contemplate what those computers will be like, which is just the same as trying to ask Galileo to imagine what his microscopes and telescopes would have looked like today. Nevertheless, if we try to look forward to the future of computation, the only instrument that can even come close is the supercomputer of today. I think that at least within 10 years from now every computer will be as powerful as the most powerful supercomputer that we have now.

Now, if we look at the supercomputer as a scientific instrument, what is it for? Well, the answer is, that it is for seeing phenomena which are not accessible to the microscope and telescope and other experimental instruments. Let me give you some examples. The standard example is tomorrow's weather, or to be more precise, the weather for the day after tomorrow. When you turn on the radio and you get the weather forecast they'll usually give you the forecast one day ahead and then they'll go on and say, well for the day after tomorrow and through the end of the week here's what we expect. That forecast has come off a supercomputer. Obviously, there is no way that the microscope and telescope, or any other such instrument can tell you what tomorrow's weather is going to be; you can't see it yet. I will come back to a bit later to how one uses a supercomputer to make those forecasts.

Another area that is not accessible to experiment is the distant past. With an telescope we can see a window. With a modern telescope, we have something like a fifty-year window, say from the Palomar telescope. Of course, the telescope really started with Galileo, so with a very simple telescope we could have four hundred years of information. But, a typical astronomical object lives for one billion years, and how do you deal with the other 999,999,600 years that we can't cover with a telescope. That calls again for a supercomputer. A major activity with supercomputers these days is to try to do modeling of the interiors of stars as they go through their lifetime. One project at Cornell is looking at the collapse of galaxies to see whether they are able to collapse into massive black holes. You can't do that with an experimental instrument, it's just not possible.

A supercomputer can be used to penetrate the very small, beyond what the microscope can see. You already know the truly extraordinary capabilities of microscopes these days in seeing atoms and, in some cases, seeing some of the details of the electrons, but mostly on surfaces. To really penetrate what individual electrons are doing, and especially the correlations between electrons, which is the basis for all materials that we work with, is still beyond the capability of microscopes. In particular, if you want to look at the properties of materials, chemicals, whatever, that we haven't even discovered yet, it is going to be very difficult to do that with a microscope.

Now supercomputers also play a very important role in applied areas. Just to give you one example, in the design of the Boeing 767 aircraft, an important design objective was high fuel efficiency. They achieved a high fuel efficiency, but the reason was that the designers were able to get some information on the flight characteristics of their aircraft designs while they were still on the drawing board. In particular, they were able to explore alternatives. For instance, to try alternative arrangements of the engines underneath the wing. What was important is they could look a lot of alternatives because of the speed of the computers they were doing the simulations on. And of course, the faster the computer, the more alternatives that they could explore in a given time. I can also tell you that one of the results of being able to explore alternatives quickly is that the engineers could go and hide and look at alternatives that would have been laughed at by their colleagues. As I understand it, that was one of secrets to the 767 design.

Now I have to tell you that the supercomputer as an instrument is young. So a lot of the things that you can imagine doing with a supercomputer we cannot do yet very well. You, of course, can see exactly how well one does with the weather because you get feedback on the order of a few days. There's no way that the results from the supercomputer can be fudged, because there's no way that the people making those predictions can know in advance what they should be predicting. And the predictions are clearly better than they were 20 years ago, but they're far from perfect.

In all the ways that the supercomputer can be used, and especially in all the ways they can be used to produce really spectacular science, there are many problems and challenges. As a result, when you look at the history of achievement of computers in terms of being used as a complete instrument in themselves, there isn't very much. There's the discovery of solitons, or to be more precise, the rediscovery of solitons because they were first discovered experimentally in the 1800's. There are the first simulations of the liquid state. It is very difficult to do analytic studies of the liquid state, because while gases can be simple, and crystals can be simple, liquids usually aren't.

You cannot really stack up the discoveries made by any kind of computers, let alone supercomputers, against the kind of discoveries that just stream out from the best of our telescopes, the best of our microscopes. But what one can see, if one has looked at this carefully, is that the opportunities for the future, both for scientific discovery and engineering applications, by continued improvements in the technology of supercomputers are vast. One cannot leave this instrument aside, even if the results today do seem more comparable to Galileo's experiments than modern-day experiments that are producing the fantastic revolutions we have today.

So first of all, I want to tell you where are some of the biggest opportunities for the future in terms of the supercomputer as a scientific instrument. Then I want to tell you something about the character of the challenge that one faces in realizing these opportunities. One of most profound opportunities for the future is simply the understanding of the chemical bond, and the ability to work with it in a precise fashion, not simply in terms of understanding the discoveries that have been made, like the high-T_c superconductors you just heard about, but to use the supercomputer to open up the extraordinarily vast area of materials, chemicals and other things that we haven't discovered yet. You only have to sit back and look at how material science has developed over the past 20 years, to look at the range of combinations that are possible among the 92 elements and the complexity of some of the materials that have already been discovered (high T_c being a good example), to realize we have only tapped an infinitesimal, infinitesimal, infinitesimal fraction of what is possible (especially given that most of the discoveries that we have so far of materials of spectacular properties have been made by accident).

Turbulence is an area which one comes up to from very many points of view in classical physics, whether a range of industrial problems such as hypersonic aircraft and how they fly, to fundamental problems such as turbulence in the plasma between the sun and the earth, or within the sun itself, or the turbulence in the universe extending up to the size scale of the universe itself. These are some of the issues that we are getting a new look at because of the simpler problem of conversion from order to chaos, the work that you heard about from Libchaber originating from the discovery of Feigenbaum.

There's the whole issue of the molecular basis of life. One's first thought is that we are learning extraordinary things about the molecular basis of life. What is the challenge? But, in fact, there is very little understanding of the way that large biological molecules, namely proteins, obtain the spatial structure, that allows them to function in living cells the way they do. It's becoming increasingly evident that that is going to be a bottleneck, if we don't come to understand that, in terms of taking advantage of many of the discoveries that we have in biology.

I've mentioned had the whole issue of the evolution of astronomical objects. Uncovering the life history of astronomical objects is again a major challenge, a challenge on which there has been a lot of progress in the past in the understanding of the sequence of the development of stars and so forth, but there is still a major amount of work to be done in the future.

Now, I'd like to say something about the character of the challenge of the supercomputer as a scientific instrument. The challenge is to move from the present situation, where there is plenty of research going on, to achieve the really spectacular discoveries which have not been made yet: the kind of discoveries that would make you run to the first colloquium on the subject, the way I used to run to the colloquia on pulsars when they were first discovered.

Let me first say something about weather forecasts. What are the problems one faces in trying to do weather forecasts? First of all, the basic problems with supercomputers in weather forecasting are rather simple. Namely, that in order to understand what tomorrow's weather is going to be, you have to describe it on the scale on which the weather is going to occur. If you want to talk about rain, you have got to describe something on the scale of a rain shower, if you want to describe a major storm you have to describe it on the scale of a major storm system. On the other hand, if you are working over several days, for example, ten days which is the European medium-range forecast, you have to be looking at the weather over the entire globe because the weather moves at something like a 1000 miles a day. If you go for 10 days, that's covering most of the globe.

To do a weather forecast you have to be able to calculate quantities like the wind velocity and the humidity. To do that you have basically to introduce a grid, or another way of looking at it, you have to divide the global atmosphere down into chunks and you have to agree that you will assign just a single wind velocity and a single humidity figure to each chunk in it's entirety. Let's look at what happens when you try to do this. Well, in today's most accurate weather forecasts what is typically done is you divide the equator into about 200 segments and then you will go from the north pole to the south pole and you divide that into about 200 segments and go from the ground to the top of the atmosphere. The people at the European medium range forecasting center told me very proudly, a year ago when I visited, that they had just moved from 16 vertical layers to 19 vertical layers. The total number of chunks requires only that you be able to multiply three numbers together. You multiply 200 times 200 times 19 and you have something like 800,000 chunks.

Furthermore, what you have to do with those chunks is to start with a snapshot of the weather as it is now, for instance at the European Center they will start with a snapshot of the weather as of noon and they have to spend between noon and 10 PM analyzing the the data (you just can't take at face value that all those balloons and weather satellites give you perfect information). They put it all together between noon and 10 PM and then they kick everybody off their Cray supercomputer and the Cray works on all four cylinders between 10 PM and 2 AM.

What is it doing? Well it has the data as best as they can work it out at 12 noon for every one of those 800,000 chunks; just numbers, wind velocity, humidity, I think there are five numbers they introduce. Then it figures out what these numbers should be for 12:15, fifteen minutes later, and it goes through every one of those 800,000 chunks and it figures out what every one of those numbers should be at 12:30. At some time about 10:30 PM it's caught up (it's figuring what the weather should be at 10:30) and by 2 AM it's gone through to 10 days into the future. Of course, this requires a huge amount of memory to keep all of that information and it requires the fastest speed that Seymour Cray has been able to produce in order that they get that forecast in time that it can be broadcast on the radio in all the European countries the next morning.

Now what is the problem with this? Well, the problem is apparent when we look at the size of one of the chunks. The equator is 24,000 miles around, you divide it into a 200 chunks, that's about 120 miles per chunk. They also showed me a map of the globe where the surface heights have been averaged to the size of the chunks they were using. This means you could not see the Alps, you could not see the high mountains of Colorado, and the highest point on earth was Greenland with that enormous icecap at 10,000 ft. They would dearly love to cut the size of their chunks in half, but that means cutting it in half around the equator. Twice as many chunks around the equator, twice as many chunks north and south. They told me they probably had enough layers vertically. As you reduce the size of the chunks, in order to have the chunk act as a unit, you can only go over a shorter time step. So they would have to go from noon to 12:07 and do something like twice as many time steps. So that means 8 times as much computing power, that means 4 times as much memory, and the computers that could do that are just coming into the market.

Now there are other problems. In trying to start from the basic laws of atmospheric science, things like the Navier-Stokes equation, they really can't get away with that because when you look carefully at what kind of grid spacing you would need to encompass all of the scales of variability in the atmosphere, that grid spacing would have to be millimeters. That's the spacing in which viscosity comes in and just keeps everything smooth enough so you can solve the Navier-Stokes equation directly with a finite difference treatment.

The grid spacing is in hundreds of miles, not in millimeters, for the actual calculation. That means they have to modify the Navier-Stokes equations in some *ad hoc* way to take into account that a hundred-mile grid spacing is not small enough to just do the straight finite-difference approximations to the Navier-Stokes equations.

They also have a stability problem. The fluctuations in the weather on a small scale grow in their scale size over time. They cannot capture those fluctuations either in the experimental data or with numerical grids. So the forecast inevitably deteriorates in quality and chiefly deteriorates in the sense that if it's accurate at all, it's only accurate in terms of it's large-scale implications. That's reflected in the weather forecasts that you hear. For example, in my home town of Ithaca the forecast for one day will distinguish Elmira from Binghamton. There's a different forecast for Binghamton and Elmira, but beyond one day it's the same forecast for everybody, because they can't make that distinction anymore. After 10 days the European forecast is sort a fuzzy global picture not even distinguishing Europe from the U.S. anymore.

Now, weather forecasting is sort of a practical thing and everybody can relate to it, but I want to discuss now a challenge of a more fundamental nature, having more fundamental implications. I want to talk about the challenge of understanding the chemical bond, electronic structure.

Now, the first thing I want to say is that the importance of applying computers to electronic structure is consistently underestimated. Because it is incredibly difficult for a scientist to imagine what it would mean if we could really do electronic structure. But let me try to give you an example, sort of an analogy, to try to give you sort of the magnitude of what we are missing both in science and technology because we cannot do electronic structure calculations.

If one wants to order up a bridge, let's say, a new bridge across the Hudson River, you can go to a design and engineering firm and you can ask them to design a bridge. They will give a price on it. They will arrange to have it built. Suppose that, instead of what actually happens with the bridge design, they had to proceed with the bridge design in the same fashion that you have to proceed if you want to discover a chemical, or a new material for building aircraft or something. If you had to approach the business of getting a bridge across the Hudson in the same fashion, what would happen is you wouldn't build one bridge across the Hudson, you would build 10 bridges across the Hudson. When they were all ready, you would have an automobile driver, a test driver, complete with parachute, in fact you'd have 10 of them, and you'd see which one of them got across the bridge.

When you look at the sociology of how chemists think, and how they work, and you compare it with how physicists think and they work, there's a profound difference. Because chemists are introduced from the beginning to problems involving bonding, which are described by the Schroedinger equation which they cannot solve, whereas physicists get their initial training on Newton's laws, which one can solve. Of course, its because of the difference between Newton's law's and the Schroedinger equation, that one has a whole capability of engineering bridges. You know if that engineering firm doesn't deliver that bridge for the hundred million dollars that it contracted for, somebody's in trouble. Whereas you cannot put anybody in trouble because they contracted for a hundred million dollars to design a new drug and didn't find one.

What is the situation in trying to apply computers to both the understanding of chemical bond and gaining ability to calculate the properties of particular materials based on their bonding characteristics? First of all, I remind you that in most quantum mechanics courses, at least to my knowledge, you learn all about hydrogen, you learn a little about helium, for instance, the variational calculation of the ground state of helium, you learn some rules of thumb about heavier atoms, and you learn basically nothing about the chemical bond. Just think back to those courses. Now why is that? The reason is the number of degrees of freedom which characterize the chemical bond.

I already described to you the problem of building grids for weather forecasting. The problem was it was in three dimensions. So I had to multiply 200 times 200 times 19. The problem would have been much easier if I only had to do two hundred grid points around the equator. That you can do on a pocket calculator.

But with the exception of the hydrogen molecule, one is dealing with grids (if you did it by grids) with three dimensions per electron. For example there's a chemical which is very heavily studied by quantum chemists, which is CH_2, that has eight electrons. That means that you have 24 dimensions. So, if you had to do a grid with a hundred points in each of those 24 dimensions, you have to multiply things 24 times. Maybe some day we will have computers which can handle 100^{24} grid points, but we're not there yet.

Now what has been happening in computational chemistry and condensed-matter physics, the two principal areas that have been working on computations of the chemical bond, is that a very rich variety of approaches have been developed. The intellectual content that has gone into some of the work in trying to study the chemical bond numerically is every bit as deserving of recognition as much of the analytic work that goes on these days. It is just that this problem is so horrible that the results have not been commensurate with the efforts that has gone into it.

Let me tell you about a few of the approaches that presently exist. The quantum chemists have been working since the early '30's. (In fact, my father started out as quantum chemist, but he abandoned it pretty quickly.) They have tried directly to extract the wave functions describing these multi-electron systems and, in particular, to get a real hold on the electron-electron correlations which are the essence of a precise understanding of the chemical bond, or precise calculations. Now they have made spectacular progress in terms of working out methods of attacking this problem of the Schroedinger equation, so that the cost, the computational cost, of solving a problem like CH_2, is far less than the cost that you have with a grid of 100^{24} grid points.

First of all, they work in terms of basis functions instead of grid points. The cost of the most accurate calculational methods these days run at something like the 7th power, not the 24th power, of the number of basis functions. Unfortunately, that spectacular reduction from 24th power to the 7th power doesn't do you very much good, because even at the 7th power, present-day supercomputers do not look very powerful. That means in real practical terms the calculations they do are limited to something like a hundred basis functions. As far as I can tell, nobody really knows how many basis functions you need to start getting real accuracy and reliability in their calculations, but it certainly isn't a hundred. If they had to go for example from a hundred to a thousand at the 7th power, that would cost them a factor of 10^7 in computing power. If you take the rate at which supercomputers are advancing, something like, very roughly speaking, 20% per year, or somewhere between a factor of 10 or a hundred per decade, then to achieve 10^7 is something like 70 years from now.

I'm certain that it will happen. There is nothing in the fundamentals of electronics, or the principles of how you design a computer, that says we can't achieve an improvement of 10^7. The only question is how long will it take? (But it is rather frustrating if you are trying to get a paper out in the next six months!)

The second method that I will mention, which again is trying to get to the fundamental accurate level of electron-electron correlations, is the so-called Monte Carlo method. This includes both a variational Monte Carlo method and a Green's function Monte Carlo method. The Green's function Monte Carlo method is specifically associated with Mal Kalos and a number of people have been working on it: Bill Lester at Berkeley, Alder and Ceperley at Livermore. I mention names here because one of the characteristics of the Monte Carlo approach is how few people have been working on it, how little real development has taken place on that subject, as opposed to quantum chemistry which has had a lot of people picking it over.

The Monte Carlo method, as you would expect, is based on producing electron trajectories, on a rather sophisticated random basis, and having the procedures by which, from those electron trajectories, one can calculate by statistical averages the energy, the ground state energy in particular, of the molecule one is studying. It has been extremely effective in things like compounds of hydrogen, including solid phases.

Most of the work has been with hydrogen, helium, and lithium in various combinations. It becomes extremely expensive as you go to heavier atoms because of difficulties with the energy scales associated with the core, the inner-most electrons. Researchers are just now beginning to try to cut down the computational cost of dealing with heavier atoms. There are plenty of ideas what on one can do, things like pseudopotentials and so forth. As I said, there are very few people and they haven't gotten around to doing it until now.

The situation with the Monte Carlo method versus the quantum chemistry method is best illustrated by the situation in the molecule lithium hydride, 4 electrons, where a very accurate calculation was done with the Green's function Monte Carlo method. It was accurate enough so it irritated the quantum chemists so that they went back and tried to do their very best calculation on lithium hydride, achieving about the same level of accuracy as the Green's function Monte Carlo. In both cases the accuracy they achieved was just barely at the level so that one could see the binding energy with reasonable accuracy by subtracting the total energy of the lithium hydride molecule from the total energy of the lithium and hydrogen atoms separately. As soon as you go beyond lithium hydride, the essence of quantum chemistry is to find the right form of black art, black magic, whatever you want to call it, so that huge errors in the absolute energy cancel out when they take that difference to get a binding energy. What happens in practice, according to people who have actually experience with this, which I don't, is that you can get things like binding energy accurately enough in about 80 percent of the times that you try these calculations for reasonably simple molecules. The frustrating thing is you never know when you're getting a wrong answer.

Let me mention briefly that there is one other method that is very popular in the physics community called the density-functional method. That is a model that treats the whole electron problem in terms of a model in which the electrons are treated as a classical fluid instead of as individual electrons with a wave function. That model has been quite successful, especially in terms of having a much larger range of application. This model has been very valuable because its range of application in present computers vastly exceeds the current limits on quantum chemistry or Monte Carlo approaches. For those of you who want to know more about electronic structure I recommend the book by A. Szabo and N. S. Ostlund on quantum chemistry (*Modern Quantum Chemistry*, McMillian, N.Y., 1982), the book by S. Lundquist and N. March on the inhomogeneous electron gas (*Theory* of *the Inhomogenous Electron Gas*, Plenum, N.Y., 1983) and the review article by Ceperley and Kalos in K. Binder's 1980 book on Monte Carlo methods (*Monte Carlo Methods in Statistical Physics*, Springer, N.Y., 1979, 1st ed., 1986, 2nd ed.)

Now, I would like to close this talk by coming back to the statement that the supercomputer is a young instrument and tell you some of the ways that it is clearly immature at the present time. I could mention the whole subject of visualization, graphics, what is in fact, the eyepiece of the supercomputer. The main trouble there, in a nut shell, is that to produce the pictures that come out of a supercomputer that allow you to actually see phenomena, the way you see into a microscope and telescope, there's a problem because with the present computing power that's behind the graphics, those pictures may come once every 24 hours. That's not very pleasant when you want to see the dynamics of what the supercomputer is telling you.

The most important problem that illustrates immaturity (and immaturity not in a bad sense, it's simply just that we're just trying to get started here) is the language of computational science. The language called Fortran. As I mentioned in the beginning, you look pretty stupid when you are actually sitting down there writing Fortran instructions. What is the problem with Fortran? The problem with Fortran is that you're trying to describe complex scientific ideas, to describe them both so the computer can carry them out, but even more importantly, to describe them to other scientists, who are trying to figure out what you did with your Fortran program, in a language which has no more capability of abstraction than the Roman numeral system.

Now, just in this past year or two, there are beginning to be programming languages which have a level of abstraction that is more like the language of algebra. I have not actually used ADA, but that is one of the languages. There is a language called C++, which is an extension of the C language. The interesting thing is that I know several scientists who have actually tried to use C++, including Peter Lepage, whom some of you probably know. Peter Lepage has been working for over a year now to try to transform his programming framework from Fortran to C++. The abstractions that are available in C++ and not available in either Fortran or C are such that he has spent over a year just trying to learn how to do it, how to actually write programs that work and do things with the capability of abstraction of C++. It will come to a real revelation to many of you to understand after 20 years of Fortran how bad Fortran really is. Thank you.

MILDRED S. DRESSELHAUS:

I'd like to thank you Professor Wilson for a most inspirational talk. Also, the other speakers of the morning. For the young students in the audience, let me translate a few of my own impressions about the messages of the morning. We are today, of course, in a golden age of science. We heard about many developments taking place in real time now, but I think another message that we also heard is about all the unanswered questions. There are so many things yet for you people to do, and we don't know the bounds of all this new discovery that will come in the next decade and beyond.

In closing, I'd like to thank the people that did all the work to make this particular symposium come to fruition, Bill Fickinger and Ken Kowalski, and also, of course, to Dorothy Humel Hovorka, who did the implementation, and provided us with the wherewithal to make it all happen.

□□

WILLIAM FICKINGER:

Now we'll introduce our Chairman. We are pleased that Maurice Goldhaber could come and chair the session this afternoon. Maurice was Director of Brookhaven National Laboratory from 1961 to 1973. He's currently Distinguished Scientist at Brookhaven. He's a recipient of the Bonner Prize and former President of the APS. He has worked in nuclear photoelectric effect, in nuclear isomers, and in the measurement of the neutrino helicity. I looked up his name in the back of Segre's book; there were references from both the front end, the nuclear part, and the particle end. Currently he's working on the proton decay experiment. Maurice.

MAURICE GOLDHABER:

Welcome to the supersession this afternoon. The year 1987 will perhaps be remembered as a super year for physics. There were supernovas and superconductors, and all the other supersubjects you shall hear about today have made progress during the year. If you look back at what happened in the 25 years following 1887, if we should have anything like so many revolutions in physics in the coming 25 years, it will remain a very exciting field. At lunch I was reminded by Roger Clapp that Michelson was famous first for developing the standard meter. He was finding it too demeaning to have to go to Paris to find out what a meter was like and he wanted to develop a meter that could be sent by telegram anywhere else.

Well, we are going to start today by hearing from Leon Lederman, who is at present the Director of the Fermilab. He previously was Professor of Physics at Columbia University, and he got experience by directing Nevis Lab first, and by fighting with the Director of Brookhaven [ed. note: i.e. Goldhaber!].

Leon is known for his many great discoveries with which his name is connected, usually he was not alone, but he was certainly a leading member each time. Among those discoveries are the K-long, the second neutrino, and ultimately, at Fermilab, the bottom quark or, as he called it, the upsilon. Today, Leon will talk to us about "The Super Collider, Assault on the Summit."

The Supercollider: Assault on the Summit

Leon Lederman

Fermi National Accelerator Laboratory

Batavia, Illinois 60510

It's very gratifying to know, whenever I see Maurice, that there is life after being a Director. I must say I'm very pleased to be part of this symposium. The only problem was the talks this morning were so elegant that the organizers have to learn never to serve the soup after the dessert.

The supercollider is the thing I will talk about, but I should comment on the morning's talks while I'm up here. This has to do with our community, this is a party for physicists. We celebrate many things. We celebrate when Professor X becomes 60 years old, and then when he becomes 65, and we give a real party when he manages to get to 70. We celebrate the birthdays of particles, like antiprotons and pions. We also celebrate discoveries. This is good because it reminds us that we are a community. I love to watch physicists display their humility, like Ivar Giaever disposing of biology, telling them exactly what to do. I was just a little disappointed that he admitted that life is complicated. Then, we heard Ken Wilson dispose of chemistry and tell them what to do. That was another aspect of arrogance. Now you're going to hear about a high energy physicist who wants to spend four billion dollars.

Arrogance is something that people complain about and I was worried about it. I told our physicists at Fermilab we should eschew arrogance. I heard one theoretical physicist pray: "dear Lord, forgive me for my sins of arrogance -- and Lord, by arrogance I mean the following...."

Collider a must for U.S., physicist warns

State offers dowry for super collider

Here's how businesses can help land crucial SSC

Superconducting Supercollider Gathers Widespread Backing

At hearings in Congress, politicians and scientists voice support for giant accelerator; some question the $4.4-billion price tag

States go after Super Collider

Illinois lies low in House hearing

Big benefits expected from SSC

Something really super

Flat land key to SSC sit

FIG. 1. SSC Headlines.

Okay, I'm going divide my talk, if I ever get to it, into something about the hype, the driving questions, and just what is the SSC. I'm going to claim that, of all the talks given in this symposium, this one is the only one which has an almost chilling relevance to the Michelson-Morley experiment, and I'll try to make that convincing to you. Then I'll talk about the technical features and wind up with some rousing conclusions.

Newspapers, preprints and articles have been flooding in. "Supercollider clears the first DOE hurdle." That's from Mississippi; this is from California, "the public is involved." Figure 1 is a composite of headlines. The amount of publicity has been simply astonishing.

I think this tremendous coverage is important, but we won't know about it for quite some time. It is a little bit like the New York blackout: the effects of it were not known for at least nine months. In 30 or 40 years from now, scientists will be photographed on their way to Stockholm and they'll be asked the usual question, "how did you get into physics?" I'll bet the number who say they read about the Collider as kids will be significant. Because, you know, billions and billions of kids are reading about SSC and, incidentally, all the other SUPER events of this year.

When the Mayor of Waxahachee, Texas can be quoted in the newspaper as saying, "we must find the Higgs boson," something is happening in America. I was actually driving through this place, and arguing with my wife as to how you pronounce it. We stopped for a bite, and we kept arguing and I asked a waitress, "could you please say, very clearly and very slowly, where we are?" And she said: " Buur-gerr Ki-ingg."

This January the President of the United States put into his fiscal '88 budget submission the proposal that the U.S. construct a 40 trillion volt Superconducting Supercollider. The price estimate was 4.4 billion dollars in current year dollars. The schedule was to have it ready in 1996. It has been called the largest basic research project ever. The step came after a large number of years, five years at least, of discussions, workshops, arguments and salesmanship of various kinds, as you can imagine. Some 25 states have spent enormous sums of money just to write the proposals in order to be selected for the site. Of course, the media hype has that as a large component. Some 230 -- I've heard it's 250 -- Congressmen have co-sponsored a resolution that this thing be authorized. Expressions of interest have been received, both on the scientist-to-scientist level but even higher, from various countries around the world to have participation in this project. So let's try to review why all of this is happening. I'm very conscious of the large number of students in this audience, which helps.

The Superconducting Supercollider is a tool for studying elementary particle physics, or sometimes high-energy physics. You heard from Ken Wilson about the telescope and the microscope. Very much this is an extension of a telescope-microscope that goes back four hundred years. Now high-energy physics is a unique subject in one sense only, in that it really doesn't have an infinite horizon as does chemistry or biology or condensed-matter physics where the number of problems, as we saw, especially from Ken Wilson's talk, are really infinite. High-energy physics will stop as soon as it answers the question: how does the universe work? When that is answered, we all promise that we'll go away, stop spending money, and do something probably more useful.

In the earliest scientific thinking, the heavens were the first object of speculation. They presented all kinds of regularities which demanded explanation. I think you can say that science, in the sense we know it now, as a search for a logical explanation of things, began with astronomy. By contrast, the world around us was much more complicated: air, earth and fire and water, rocks, and metals and mists, and so on, all kinds of things including the complexities of life. This was much harder, but there is a thread which combines both of these subjects, which we can trace back in kind of an amateur historianship.

I'm going to violate the commandments of the conference organizers even more than Ken Wilson, who only went back four hundred years, I'm going back 2500 years. In Miletus, which is probably where it all began, at least, if you believe Aristotle, that was where we started trying to explain things by logic, reducing the influence of myth and superstition and introducing the practice of rational criticism and debate. Here, the notion arose that the universe could be understood. In 450 B.C. came the second great breakthrough, which had the first proposal that there are such things as atoms, uncountable objects, too small to see, moving randomly in a void, and by clustering they make the things we can see and touch and smell. These primitive, primordial things are assumed to exist. So began the notion of atomism which was admired and adopted by many of the succeeding philosophers, including Galileo and Newton. However, it remained more of a philosophical vision until the 19th century chemists first established the experimental evidence for atoms.

We began this century with this great edifice of Maxwell-Newton classical physics, but with some puzzles including the Michelson-Morley experiment. This was a puzzle because we lost the ether and we lost the reference frame by means of which we could describe how light traverses space.

In 1900 the electron was discovered; this can truly be said to have started the modern phase of particle physics. Already with the electron the question was raised (and it is still with us) which is even relevant to the SSC. It was a comment of Lorentz's, that he didn't understand the electron because God makes electrons and in assembling the electron one had to squeeze a lot of charge together in a very small space. This took a lot of work and the work should have shown up in the mass of the electron, but obviously it didn't because the mass of the electron was too small.

In 1910 we have the discovery of the nucleus in some sense exacerbating the great puzzles that were beginning to torment the classical theory. In 1905 we had relativity. In 1925 we were beginning to see the quantum mechanics in its full glory. Later, the masters of the quantum mechanics went on to study the nucleus and we had the discovery of the strong interactions. In the 1940's quantum electrodynamics flowered. Then we went from pions to quarks in the next fifteen years. In the 1970's, electroweak unification, the Higgs mechanism, quantum chromodynamics; in the 80's, what we call the standard model, grand unifications, sort of a new unity of particles and cosmology, and questions raised, like what is the electron, what does it really mean. Perhaps by the year 2000, if we play our cards right, we will have solved all of these problems.

Indicative of the state of physics of the 1900's, in the beginning of the century, was the periodic table of the elements, which was the basis of empirical chemistry; but there was no understanding at that time of the regularities. In fact, Mendeleev had a hard time selling this table because people were very suspicious of numerology. Prout had almost sold his crazy theory that everything was made of hydrogen; then when experiments got more accurate it was clear that what we now know was a great idea, just didn't seem to work. Mendeleev bore the brunt, but the table worked as an empirical table for things. Then, of course, this gave rise to quantum mechanics.

In some similar way, a table of the 1980's shows a list of particles that have been measured. This is a triumph of experimental science because the vast majority of these hundred or so particles live for times of the order of 10-20 seconds. Yet their masses were determined, and a wide variety of their quantum numbers, and these particles were duly entered into the various tables. The period of time, in the 1960's, was very depressing, because the simplicity that had been promised us by the Greeks, that there must be an overarching, simple scheme which explains the universe, looked pretty remote. Well, like biologists, physicists decided to classify. Most of this work was done by the next speaker [Professor Gell-Mann], so I won't go into too much detail. Organizations of the particles, such as those sketched in Fig. 2, were made that were suggestive, but still we had this great puzzle of the complexity.

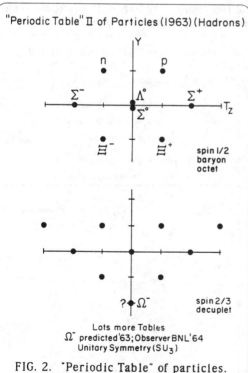

FIG. 2. "Periodic Table" of particles. Baryon octet and decuplet.

Now, I want to discuss the experimental aspect of the subject. In more recent times, an enormous number of inventions of great ingenuity and an enormous investment in time, effort, and public funds went into the machines which gave rise to the data which were accumulated. Strong focusing magnets, intense beams of neutrinos, colliding-beam accelerators, invention of antiproton cooling, the development of high-field superconducting accelerator magnets in 1980's, all of these contributed in the accelerator art. Bubble chambers, more sophisticated electronic devices which rendered trajectories visible with nanosecond time resolution and micron space resolution, faster digital electronics for data acquisition and data processing similarly contributed to the detector art. Combining these, we had the late 20th century version of Galileo's telescope or

Hooke's microscope giving us a new perception of objective reality.

Well, like the periodic table of the elements, the pattern of created sub-nuclear particles suggested that the particles had an internal structure. The proposal was that three unseen particles ("quarks") combined to make up all the entries listed in the tables of elementary particles. They are now given the names up, down and strange. You write drugstore prescriptions for how to make up the neutrons, protons, lambdas and the other things. You attribute properties to the quarks which are delicately designed to make up the particles that are seen. This was a tremendous synthesis.

Alas, three quarks weren't enough. In the parallel lepton sector (the particles that do not have strong interactions) there was an electron, a neutrino and a muon, and this was augmented in 1961 with another neutrino. Collectively these objects were called leptons. That suggested the possibility, via an intuitively held feeling for the symmetry between quarks and leptons, that there be another quark. Sure enough it was found in 1975 and called the charm quark. Then a little bit later a fifth lepton (tau) was found and then a fifth quark was found. The implication is since all the other charged leptons have neutrinos, the tau lepton must have a neutrino, otherwise it wouldn't be fair. Since the fifth quark was called bottom, there had to be a top. But these, of course, came from the symmetry of the thing, so everyone now believes that there are six quarks and six leptons. One quark still has to be found called the top, and one lepton still has to be seen, although the indirect evidence for its existence is pretty good.

Well, a vast amount of data has been acquired on the detailed behavior of the strong, weak and electromagnetic forces. Of course, they don't only push and pull, but they produce changes of species. Then it turned out, if you listen carefully to theorists, the only consistent mathematical description, in accordance with the data that was presented, is a quantum field theory exhibiting a particularly important kind of symmetry called gauge symmetry. Gauge symmetry implied that the forces are transmitted through space by particles which are called gauge bosons, if you like, the quanta of the force fields. So, electromagnetic forces are carried by photons, weak forces by W's and Z's (massive particles recently discovered at CERN) and strong forces by gluons. The periodic table of the elements as of now, which is also called the standard model, or

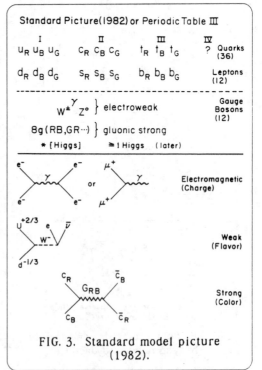

FIG. 3. Standard model picture (1982).

been cooling. At the earliest moments matter was decomposed into its most primordial components and presumably that's where the connection is between particles and cosmology. To model the universe, one has to understand the primordial components. So, there's a two-way flow of information. In fact, the big bang theory provides the ultimate accelerator laboratory (with an unconstrained budget, somebody said) and gives us a sort of terminus of how physics can progress.

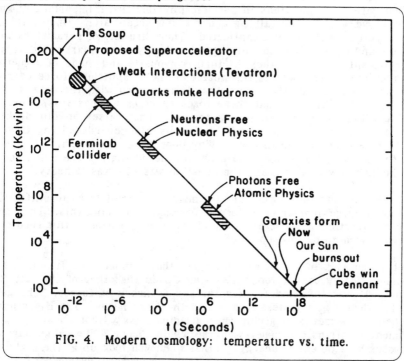

FIG. 4. Modern cosmology: temperature vs. time.

In Fig. 4 we plot temperature against time. If you know the physics you can now trace the universe backwards from the present, to galaxy formation, to atomic physics, to nuclear physics, and back to where the temperature of the collisions in the supercollider will replicate the early universe at around 10^{-16} or 10^{-17} seconds. Somewhere near the right side the Dow Jones is going to cross 4000 and the Chicago Cubs will win the pennant.

Cosmology has open questions too. What is very exciting is that something very much like the Higgs phenomenon appears in one of its open questions. Namely, in the inflationary version of big-bang theory, there appears a long-range scalar field. Inflation explains many, many things that were puzzling about the big-bang theory, but then presents a new phenomenon, a new field which complicates the vacuum and has properties that are very reminiscent of the Higgs particles that appeared in particle physics. So, the particle physics puzzlement described by Higgs has a parallel in cosmology and that seems to us to be very exciting.

The standard model of particle physics is very powerful; it explains all the standard picture, is sketched in Fig. 3. It consists of six quarks which come in three colors (color being a quantity analogous to charge for the strong interactions), six leptons, and they're carried by 12 gauge bosons -- the photon, the W and the Z, and the gluons. Then there's something else called Higgs, which I will say a few words about, because that ties us into the Michelson-Morley experiment.

There are open questions in particle physics; that's important to know, otherwise we wouldn't be talking about a new accelerator. One of them is that it's getting a little bit complicated. There are lots of parameters which we don't understand how to put into the theory. Are quarks and leptons truly elementary, or are they in turn, composite and made of something even simpler? Then there is the puzzle of the W and the Z force carriers, which are massive, but the theory that is in accord with the data, i.e. the gauge symmetry, insists that these force carriers be massless, otherwise terrible things will happen to the theory. That raises an old problem, where does the mass come from? A famous blackboard writing ("Do not erase!") of Richard Feynman's was "Why does the muon weigh?" I. I. Rabi, when the muon was discovered said "Who ordered that?" The muon seemed to be just a heavy electron, and the question was why was it heavy.

We believe that four forces are too many. We want to go to unification. There are a large number of different proposals for doing this. They go in different directions. These are some of the open questions. Our theoretical colleagues seem to be in deep trouble.

Let me go back to this Higgs thing; not that I understand it, but it's the Michelson-Morley connection. It's often called the ether of the late 20th century. It was invented to save the gauge theory of forces. The gauge particles should be massless; the W and the Z are heavy. The Higgs potential then was a means of giving the particles mass without destroying the fundamental intrinsic symmetry of the whole theory. The gauge particles get masses by a potential energy. So that when you look up at the night sky, not only do you see the stars, but out there somewhere, everywhere, is this potential in the vacuum which changes the vacuum. It is the Higgs potential.

It turns out that all known masses must be explained in terms of this potential energy of the Higgs in the vacuum and, as such, the Higgs field is nothing but the good old ether in thin disguise. It's not needed to define a frame of reference as the 19th century ether is, but it's needed to provide a basis for potential energy. Now, the point is then, that the SSC, which is specifically designed to test this idea as anything else, then becomes the Michelson-Morley experiment, in a philosophically very analogous way, a hundred years displaced.

Let me just say a few words about telescopes instead of microscopes. Our colleagues in astrophysics and astronomy have made enormous progress. Again with equipment of fantastic ingenuity, opening up the electromagnetic spectrum, through radio, gamma rays, and most recently neutrinos. There is a standard model of cosmology which says that the universe started about 15 billion years ago in a hot big bang. Since then the universe has the data we have to date. There is no data known today inconsistent with the standard model. Consistency of this powerful theory demands a Higgs field and we are able to predict something about these Higgs fields and Higgs particles. In particular, the standard model is able to assure us that the mass of the Higgs particle must be on the order of, or less than, 1 TeV. This important conclusion follows from what are generally classified as unitarity arguments. If the mass of the Higgs is heavier than about 1 TeV either terrible things happen to the theory or new phenomena must appear. If the Higgs idea is wrong, something else must take its place which will also be new physics. It might be a new strong interaction among W particles or something else. In this sense, somebody called the SSC a no-lose proposition: some new physics has to appear before the mass scale of the order of 1 TeV in order for our present understanding of the micro-world to be consistent.

How do we insure that we can study the 1 TeV mass scale? We have to re-view the nature of particle collisions. It turns out that protons are by far the least expensive way to get the very high energies. We know how to do it, but a proton is a complex object, sometimes compared to a garbage can. It has quarks in it, two up quarks and a down quark; it has gluons in it, and those gluons themselves can reappear as heavier quarks. It's a very com-plicated object, and when two of these things collide there is a lot of debris. And when we want to get 1 TeV excitation, we mean 1 TeV on the constitu-ent level; that means one must divide the incident energies by some factor that has to do with the sharing of energy among the constituents. It turns out that a "rule-of-thumb" average factor is somewhere between 10 and 20, although we know how to compute the factor very well for specific colli-sion processes.

So, in brief summary, the ether search in 1887 was the Michelson-Morley experiment; I claim the analogous ether search in 1997 will be the su-percollider. What is it? Two concentric rings, each accelerates and stores protons at 20 TeV. They make head-on collisions, giving a total energy of 40 TeV. One must then derate that by these constituent factors. This gives us 2-4 TeV on the constituent level. The luminosity, that means the number of collisions per second, is such that you can actually look at, or decide to look at, some fraction of 10^8 collisions per second taking place. Colliders al-ways have the penalty that hitting protons against protons is a rare pro-cess, like hitting machine gun bullets against machine gun bullets. Thus, getting to 10^8 is really a great achievement. The magnets are supercon-ducting and the superconducting magnet technology is based on niobium titanium, which is an ancient technology they tell me, but it does give 6.6 Tesla, 66,000 Gauss, magnets. That makes the tunnel about 80 kilometers in circumference and that is quite a large object.

I'd like to give you a glimpse of some of the theoretical estimates which supplement the idea of simply dividing the 40 TeV by 10 or 20. Many calcu-lations were done using various models for what the future may look like. The key idea though is that we do know matter consists of quarks, and the fundamental collision will be a collision between two quarks or between a quark and a gluon. One does know a lot about the motion of the quarks and gluons inside the nucleons; that's the bread and butter of today's experi-

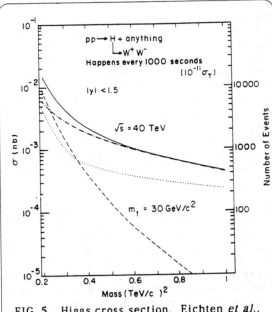

FIG. 5. Higgs cross section. Eichten *et al.*, *Rev. Mod. Phys.* **56**, 579 (1984).

ments. By knowing the distributions and convoluting the process with the proposed cross section, one can do very reliable estimates of the cross sections to be expected, for example, for the Higgs. So, there are tables of these things that are available. A whole issue of *Reviews of Modern Physics* is devoted to this. An example is shown in Fig. 5, where the predicted cross section for W^+W^- pairs is shown. One can look up the mass of the Higgs, since we don't know it, and look up the number of events you get for the given machine and for the given energy. It was on this basis and many graphs like this that one designed the SSC parameters.

Of course the Higgs isn't the whole story. The machine will open up a vast area of unknown territory and parameter space to explore. We may expect to learn something about heavy quarks and heavy leptons, and more gauge bosons, and questions of supersymmetric particles and technicolor. We will be able to look much more incisively at whether or not the quark is a point object or, in fact, itself composite. All these things are well enough along in their speculative proposals to allow a book of calculations on the basis of which experiments can be designed.

Let me then show you what the SSC looks like. In Fig. 6 is a drawing of what the machine might look like. It's elongated with the straight sections very long in order to allow a series of non-interfering experiments. The experiments are carried out at the straight sections. The rest is just a ring to keep the protons going in a circle so you can keep accelerating them across the radio frequency gap. The way accelerators work now, as the accelerated particle speeds up, you have to shift gears, from low gear to high gear.

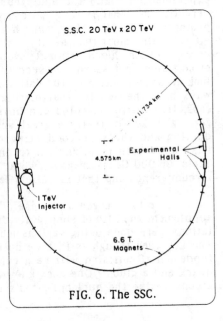

FIG. 6. The SSC.

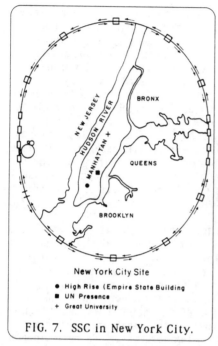

New York City Site

● High Rise (Empire State Building
■ UN Presence
+ Great University

FIG. 7. SSC in New York City.

That's done by a series of pre-accelerators of which the largest is a 1 TeV accelerator, the small circle shown in the figure. That 1 TeV accelerator then injects into the big ring, and there are smaller accelerators which get the protons into the 1 TeV accelerator.

I was hoping to get a transparency for what the machine would look like in Ohio, but it never arrived. So I used my favorite site which surrounds New York City, as shown in Fig. 7. New York City has lots of advantages. There's already a high-rise and an administration headquarters (City Hall), there's the United Nations Building, and there are even a lot of tunnels, if you could only move them around. So here you see again the scale of the thing. Of course, the vast portion of the machine will be underground, probably fairly deeply underground, and will stay there and not make trouble and be much more or less like a gas line, or a water pipe, with occasional access shafts for bringing down the utilities; it will be inert and benign. Normal surface activity can take place, but there will be places, campuses if you like, where the experiments are done. Those will be fairly extensive and well developed. Figure 8 shows you a cutaway of the tunnel.

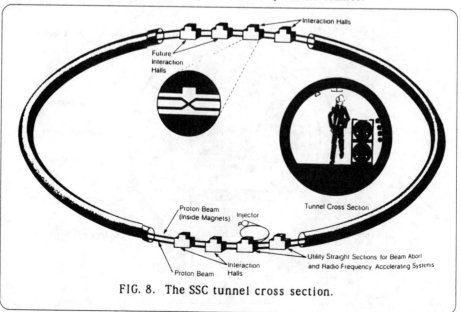

FIG. 8. The SSC tunnel cross section.

The magnets themselves are things of enormous complexity and have
been in a development stage, I think, since the invention of hard superconductors in the 1960's.   At that time, high-energy physicists seized on
this discovery and first built instrumentation magnets for bubble chambers and spectrometers. They then began the research, almost world-wide,
for how you make ramped magnets required for accelerators. This is very
much an AC magnet.   The magnetic field has to increase from a fraction of
a Tesla to 6.6 Tesla, in a certain period of time. The magnetic field itself has
to be precise in all of its multipole moments, which go out to about the 20th,
to about one part in 10^4, but sometimes even better. That's because the particles stay in this vacuum and go around the eighty-kilometer circuit some
four thousand times a second and are supposed to stay in this machine for
as long as one day and never depart from their prescribed orbits by more
than a few millimeters. The precision on the control of particles is largely
given by the precision of the magnetic field.

In the superconducting machine the magnetic field is determined by the
position of the current-carrying elements. Now, think of the fact that you
wind the machine at room temperature and then you cool it down to four
degrees and things shrink by inches.   Then when the magnetic field is
turned on the forces on the current elements are enormous, thousands of
pounds per square inch! Nevertheless, the current elements must maintain
their design position to a few thousandths of an inch.   This should give
some feeling for the engineering challenge on the magnets.   The magnet
itself, of course, is cooled to liquid helium temperatures. To restrain the
forces, not only is there a very strong stainless collar, but also a large iron
collar. This serves also as a magnetic return path but plays little role in the

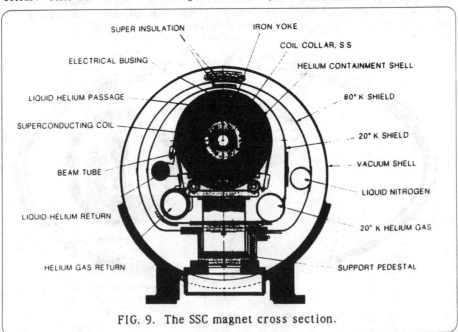

FIG. 9. The SSC magnet cross section.

shaping of the magnetic field at the orbit. The magnet structure is placed into a thermos bottle (cryostat) to keep the cool, if you like. A cross section is shown in Fig. 9.

It turns out that most of the cost of making a magnet is in the ends. If you'd like to reduce the number of ends, you make the magnets longer. Of course, there's a limit because the stored energy in the magnet must be dissipated harmlessly in case of trouble and the longer the magnet is, the more stored energy there is. So, there are all sorts of compromises and the currently proposed magnet length is about 16 meters. There are eight thousand of these magnets that will have to be produced. There will also be a smaller number of shorter quadrupole magnets needed for focusing the beam.

Let me review, just quickly, some of the major technical and engineering issues. The field is very high, 6.6 Tesla. The prototype of the SSC is the Fermilab Tevatron which runs with magnets at 4.2 Tesla and with superconducting materials that are a lot poorer than the materials that are now available. The improvement in niobium titanium metallurgy has been very significant, and at the same costs one can get over 1.5 times the magnetic field that one had in 1980 when we began the Fermilab Tevatron.

I've already mentioned the precision requirements, the placement stability, the ramping from 0.3 to 6.6 Tesla, the enormous forces that are applied to the conductor that must be resisted, the need for fine wires to reduce the inherent effects of residual currents, the specifications on materials of almost 3,000 amps per square millimeter at a field of 6 Tesla. Then there is quench protection, the notion that you have an enormous amount of stored energy, over a megajoule per magnet (about a stick of dynamite) and you'd like not to have that go off if you can avoid it. You need a very, very good vacuum, less than 10^{-9} Torr, because the particles are being stored for a such a long time and collisions with residual gas atoms would destroy beam quality and reduce storage time.

The controls and diagnostics must reach a new level of sophistication because of the scale of the project. Then, of course, particle detection problems become significant. At Fermilab we're coming to grips with analyzing data for 2 TeV collisions. These are complex and formidable. We think that the collisions are subject to analysis. In fact some of the data we're beginning to see at Fermilab are beautiful. However, getting up to 10^8 collisions per second, where you only have ten nanoseconds to digest an event before the next one arrives, that's a major challenge. In spite of heroic efforts we must still wait, and I believe for the next 5 or so years for a good solution to that problem.

There must be an enormous data acquisition system, what amounts to a supercomputer in the sense of Kenneth Wilson's talk, because what you have is 10^8 collisions, at something like 10^6 bytes per collision. So you're dealing with 10^{14} bytes per second, and you've somehow got to find the golden events, which probably occur at maybe a few events per second. You've got to filter the data very rapidly through a whole sequence of levels, until you've finally reduced the sample to the few events to store in order to examine at greater leisure.

There is a catch to all of this and that has to do with the cost. The cost is impressive. Fermilab was the largest accelerator of its time, it was built between 1967 and 1972. Its cost, in current dollars, including the initial complement of detectors was about 1.5 billion dollars, but the SSC will be about three times higher.

	FY 86 K$	FY 88
Superconducting Super Collider	3,010,318	3.21
Technical Components	1,424,161	
Injector systems	189,252	
Collider ring systems	1,234,909 → Cryo:	120M
Conventional facilities	576,265	
Site and infrastructure	65,433	
Campus area	42,860	
Injector facilities	39,758	
Collider facilities	346,803 → Tunnel	
Experimental facilities	61,412	
Systems engineering and design	287,607	
EDI	195,404	
AE/CM services	92,203	
Management and support	192,334	
Project management	114,749	
Support equipment	52,635	
Support facilities	24,950	
Contingency	529,951	

Other Costs	
R&D in support of construction	275 M
Detectors and Computers	720
Pre-Operating	170
	$ 1165 M

FIG. 10. SSC cost summary.

The cost, 4.4 billion dollars, includes a reasonable allowance for detectors, as far as you can estimate them. This cost has been studied very carefully by an insistent bureaucracy and probably is an accurate cost estimate. It's an enormous sum of money and creates all kinds of socio-, political-, science- policy issues.

I might say a little about international collaboration because one of the hopes is to reduce the cost, at least to the U.S. taxpayer, by international collaboration. High-energy physics carries on a long tradition of being good on international collaboration. CERN is probably the quintessential example of a mode of collaboration where many countries chip in to buy the equipment and support the laboratory. Fermilab, equally international, has a different mode. It's used by scientists from twenty countries with formal and informal agreements, but contributions come only through the construction of detectors. The operations are paid by the host Lab.

In West Germany, where they're building an impressive new machine, they've struck a balance between these two modes of collaboration. Four countries are contributing to the construction of the accelerator, in one-on-one agreements. This is the model we're looking at in order to try to interest various other countries. All of this is still in a highly tentative state since the machine has not yet been approved by the Congress.

Let me wind up with one or two summary remarks. I think there's a consensus that the inventory of accelerators that we have will not be capable of addressing these most profound issues and the SSC is proposed as a remedy. The goal, as I said, is a 20-on-20 TeV proton-proton collider. If it appears in 1996, it will represent a 13-year planning cycle, maybe longer.

One of the dominating motivations (certainly not the exclusive one and this is a subjective opinion) has to do with this Higgs business, sometimes called a Higgs sector, or a Higgs system. One of my favorite theorists said it truly serves as the collective of the main problems of high-energy physics; it embodies all of our ignorance of structure of elementary particles, because of its connection with mass. This reminds one of Lorentz's comment on the electron.

Concluding, why should we build this thing? Well, first, there's no substitute for addressing those particular issues, which I call the 1 TeV mass scale, in this time period. We looked at all different kinds of alternatives. LHC is a proposal of our European colleagues. Linear colliders are another kind of accelerator which requires at least four or five years of continued R&D and one is not certain how that will come out. People have talked about using high-T_c superconductors; that can be confidently ruled out on the scale of the next decade. Nor can you, in spite of Ken Wilson's optimism, do it with Monte Carlo programs, because you need the input data, and that's the problem.

I think you build it in order to continue the quest to understand the universe. You don't build it for spin-offs; that would be a dumb thing to do. It has spin-offs, these are beneficial, they reduce the overall cost to the Treasury, etc., but that's not a reason for doing anything. You don't build it for national supremacy. It probably won't directly produce world peace or insure economic competitiveness, which are other arguments I've seen. I do know that while the Senate committee was considering the SSC and its costs, one fellow said we ought to authorize 35 million dollars, just at that moment the stock market crashed. I don't think we can use that as an argument pro or con.

I think you do it to support good science. We live in a great nation; great nations should support this kind of bold initiative, the continuation of an age-old quest to understand nature. And with all the hype, the four supers that we talked about, there's never been a more positive public attitude towards science, in my impression, than now. Thank you.

MAURICE GOLDHABER:

Thank you, Leon.

□□

MAURICE GOLDHABER:

Thank you, Leon. We now come to the next "Super"-man. Murray Gell-Mann is at present Millikan Professor of Theoretical Physics at Cal Tech. He got his PhD at MIT, and he became known immediately for his deep insight into the structure of particles. He was the man who introduced us to quarks and their colors, and he will talk to us today about superstrings.

Is the Whole Universe Composed
of Superstrings?

Murray Gell-Mann

CALIFORNIA INSTITUTE OF TECHNOLOGY

Pasadena, California 91125

Thank you very much, Maurice. The great archaeologist, Sir Arthur Evans, after he retired at the end of the 1930's from digging in Crete, revisited some of the places that he had known in his youth in what is now Yugoslavia. In fact, it was his noticing certain seals in the Balkans that led him on the quest toward northern and southern Greece, and finally to Crete to pursue the civilization that had made those seals. Revisiting the Balkans in his old age, he went to a fortress where he had been imprisoned as a young war correspondent. The guide showed them around the prison and then offered to take the them to the dungeon. Arthur Evans strode ahead of the guide, who said, "Excuse me sir, have you been here before?" Sir Arthur said, proudly, "I come here every fifty years".

Well, I come to Cleveland every 35 years and it's a pleasure to be back. When I was a student at Yale I remember hearing that we in Connecticut had a sort of colony out here called the Western Reserve, and I'm glad to hear that there have now been some significant achievements in science here in the Western Reserve, including a disproof of the existence of the ether. We can now discuss some of the remote consequences of that work.

You've already heard a kind of introduction to our field of particle physics from the previous speaker. I remember when he was one of the students in my class at Columbia. He was an above average student. It's nice to know that he's kept up in a sort of vague way with the developments in the theory of the elementary particles.

We study the fundamental building blocks of which the whole universe, including us, is made. Each kind of particle has the same properties wherever found anywhere in the universe. The laws of those particles and their interactions constitute the fundamental microscopic laws of natural science. The macroscopic laws concern the boundary condition in the early moments of the expansion of the universe, especially the fact that some 15 or 20 billion years ago it was a tiny, dense, hot, expanding ball. The two subjects, elementary particle physics and early cosmology, have practically merged, and nowadays many of the most important ideas in cosmology are coming from elementary particle physics. Together, the two subjects constitute the fundamental laws of natural science, in the small and in the large. They underlie all of physics and chemistry, and in turn they form the substrate for the rest of natural science.

Ed. note: The title of this talk in the program of the symposium was "Superstrings."

This article discusses theory. There isn't room for more than a glance at experiment or observations, but we must remember that theoretical science and experimental or observational science have to advance as partners. Sometimes theory is ahead and predictions are confirmed by observation. At other times experimentalists find something unexpected and theorists have to go back to the drawing board. (One example, to which we shall refer briefly later on, is the discovery of the tau lepton and then of the **b** or " bottom" quark, the existence of which was demonstrated by Leon Lederman and his collaborators when they detected the "upsilon" meson, after a false start that is known as the oops-Leon.)

Our subject, superstring theory, to which we will get after a long intro-duction, has so far very little basis for comparison with experiment, al-though it is fantastically promising as a theory, as I shall try to make clear. We must remember that ultimately observation and experiment will be the arbiters of its validity.

I shall start by reviewing some aspects of elementary particle theory that are relatively secure, confirmed in large part by experiment. When we get to more speculative material, I shall issue an appropriate warning, which is necessary because both kinds of theoretical ideas sound equally crazy.

All of modern theoretical physics is based on a few principles. The most important of them is that mysterious discipline called quantum mechanics, which was invented some sixty years ago. It is not a specific theory, but rather the framework within which all correct theories have to fit. No-body feels perfectly comfortable with it, because it is what the social scien-tists like to call counter- intuitive, but we know how to use it, and it works perfectly.

Einstein's principle of relativity, which grew out of the Michelson-Morley experiment, also seems to be perfectly correct, and so does causality, the very simple principle that a cause comes before its effect. If we put these three together we get the fundamental framework of quantum field theory, within which any respectable work or even speculation in our field has to be carried out.

In quantum field theory every force is carried by a particle that is the quantum of that force. For electromagnetism, the quantum is the famous photon, very well confirmed by experiments since it was suggested in 1900. We have a beautiful theory, quantum electrodynamics, which describes electrons and their interactions with photons. It is now about 58 years old and it is a perfect model of a successful theory, confirmed up to an accura-cy of many decimal places by observation, so that it deserves its nickname of QED. In any quantum field theory we can draw funny little pictures, in-vented by one of my colleagues, which are supposed to give the illusion of understanding what is going on in field theory. In Fig. 1 we have electrons exchanging photons to get the electromagnetic force between them. The electrons have negative electric charge, as indicated by the minus signs. (The sign convention is due to Benjamin Franklin and has no special im-

portance.)   If you know some
physics, you will note that the
emission of the photon by one
electron and its absorption by the
other seem to be forbidden by the
laws of energy and momentum
conservation; but if you know still
more physics you will understand
that in the Pickwickian sense of
quantum mechanics, a photon can
still be "virtually" exchanged because over short space-time intervals those laws do not have to be
exactly obeyed. (That is an example of the   "Heisenberg uncertainty principle" in operation.)

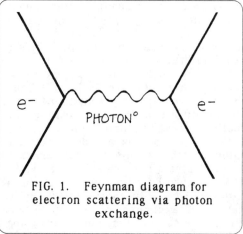

FIG. 1.   Feynman diagram for
electron scattering via photon
exchange.

The history of our field over
centuries is that we have come to understand matter in smaller and smaller
chunks. We have learned on several occasions that what we thought was
fundamental is in fact made of smaller things: molecules and crystals are
made of atoms or ions; the atoms or ions are made of nuclei with electrons
around them, as is well known; and the nuclei are made of neutrons and
protons as physicists began to understand around 1932, when the neutron
was discovered.   Now we know that the neutrons and protons are in turn
made of quarks, where "quark" is the obvious name for their fundamental
constituent.

The prescription for making a neutron and proton out quarks is that you
take three quarks.   The electically neutral neutron (electric charge zero) is
composed of one **u** quark with charge +2/3 and two **d** quarks of charge -1/3
each, giving a total charge of zero.   For the proton, the charge of which is
taken to be the fundamental unit of charge (so that its electric charge is
+1), we have two **u**'s and one **d**.   Adding up 2/3, 2/3 and -1/3, we get 1.

The **u** and **d** are called flavors of the quark, and there are actually more
flavors, although not as many as 31.   Probably there are six.   Besides flavor,
the quarks also have a very important characteristic that we call by the
nickname of color, although it has absolutely nothing to do with real color.
The name is just a joke and a metaphor.   There are three colors, and we call
them red, green and blue, after the three fundamental colors in a simple
theory of human vision.   The prescription for a neutron or proton is that
you take one quark of each color, i.e., a red quark, a green quark, and a
blue quark in such a way that the color averages out.   Since in vision,
white can be thought of as a mixture of red, green, and blue, we can use the
metaphor to say that the observed particles like the neutron and proton are
white.   Color is averaged out in the observed particles and only inside them
can colored objects exist.

Just as electrons have the electromagnetic force between them that
comes from the virtual exchange of quanta called photons, with which
many people are familiar, so the quarks are bound together by the

exchange of other quanta called gluons, with which most people are unfamiliar. These quanta "glue" the quarks together to make the neutrons and protons and other observable nuclear particles. The gluons pay no attention to flavor – we can say that they are " flavor blind." However, they are very sensitive to color and for different situations there are actually different kinds of colored gluons. In Fig. 2 we see how a red quark turns into a blue one with the virtual emission of a red-blue gluon, which is then virtually absorbed by a blue quark, turning it into a red one. In another situation we have the exchange of a blue-green gluon. We can say that gluons are colorful, and we shall see how that makes an enormous difference to the character of the theory that applies to quarks and gluons.

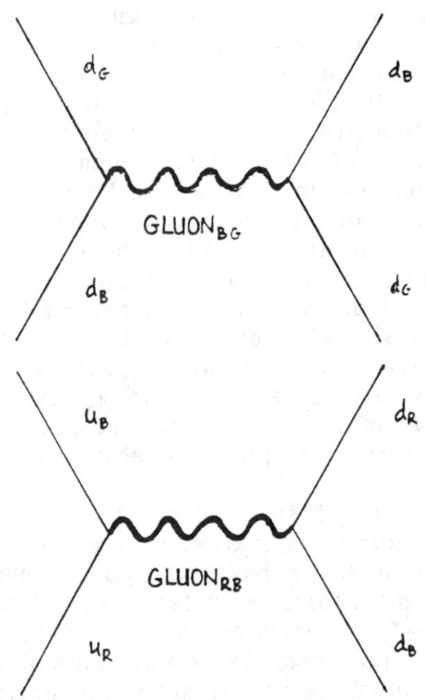

Around 1972 some of us proposed a definite quantum field theory for quarks and gluons to describe the nuclear particles and to explain the old mystery of the nuclear force. The theory is called quantum color dynamics, or quantum *chromo*dynamics if you like to use the Greek word for color. It seems to be correct, although not yet absolutely proved. The nuclear particles are all made of quarks and gluons behaving according to this detailed mathematical theory. We can put together a sort of a dictionary as illustrated in Fig. 3.

FIG. 2. Feynman diagrams for quark scattering via gluon exchange.

In quantum electrodynamics (QED), we had electrons interacting through the exchange of photons; in quantum color dynamics (QCD), we have the colored quarks of various colors interacting through the colorful gluons. The newer theory seems to be just as valid as the older one, but we still need further calculations, perhaps on one of Ken Wilson's futuristic computers, to make sure that it is really correct.

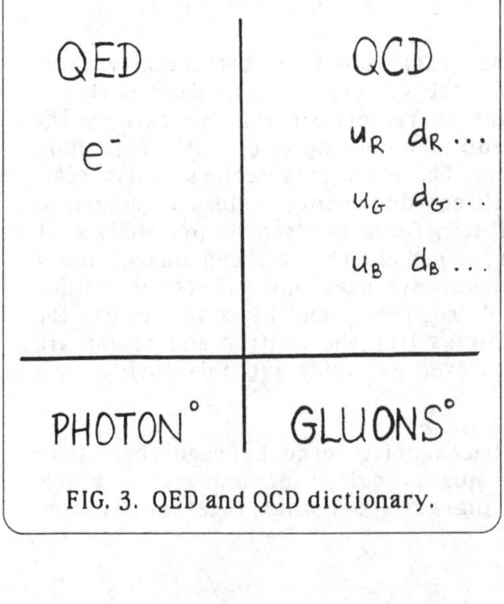

There is an important difference between the two theories. Whereas in QED, the quanta that carry the electromagnetic force are electrically neutral, the situation in QCD is different: the quanta

FIG. 3. QED and QCD dictionary.

that carry the color force are themselves colorful. As a result, the equations have a couple of extra terms, the solutions of the equations are different, and it turns out that the color force is completely different in its properties from all other known forces: it does not drop off at large distances
like gravity or electromagnetism. It stays big even up to infinite distance.
That in turn has the effect that all colored particles like quarks and gluons
are permanently trapped inside the white observable objects like the neutron and proton. The constituents cannot emerge to be detected individual-
ly, although there are numerous experiments that reveal these constituents inside.

We believe (although the relevant calculations have not been worked out
completely by any means) that the nuclear force, which holds the neutrons and protons together in the nucleus, can be understood in terms of
the basic quark-gluon interaction. We seem to have solved, at least in principle, one of the main problems that faced elementary particle physics
when I was a graduate student.

Now there is more to the world than the atomic nucleus. In an atom we
have electrons surrounding the nucleus. The electron has a charge of -1,
but it has no color and does not feel the nuclear force. (An electron inside
the nucleus, such as one of the most innermost electrons of very heavy
atoms, feels only the electrical force of the protons.) However, the electron
does in some sense have flavor. It has a kind of silent partner called the
electron neutrino, which represents, so to speak, another flavor of the
electron, just the way the u and d are different flavors of quark. The electron neutrino, which was first suggested theoretically by Wolfgang Pauli
in 1930, feels neither the nuclear force nor the electromagnetic force and
it can pass right through the earth with very little probability of interacting. The solar neutrinos, produced in the nuclear reactions in the center
of the sun come up through the earth at night.

The poet and novelist John Updike, when he read about this situation
around 1960, was inspired to write the poem, "Cosmic Gall."

(In the third line it is tempting to employ scientific license and alter "do
not" to "scarcely.") Unfortunately detection of solar neutrinos is still
fraught with many problems. The rate of detection is much lower than
predicted, leading my colleague Willy Fowler to suggest that maybe the sun
went out some time ago and the news has not reached the surface. I think
not many people believe that explanation, but if it is true, it means we are
headed for a real energy crisis.

The electron neutrino can be detected because there is a force in which it participates, along with the electron: the so-called "weak force," which allows, for example, the reaction

$$\nu_e^0 + n \rightarrow e^- + p^+,$$

or, more basically, in terms of quarks,

$$\nu_e^0 + d^{-1/3} \rightarrow e^- + u^{+2/3}.$$

The electron neutrino turns into an electron, while the **d** quark turns into a **u** quark of the same color. As indicated in Fig. 4, this process occurs through the virtual exchange of a heavy, electrically charged quantum called X^+ or X^-, which theorists (including me) predicted in 1957, and which was finally discovered experimentally in 1983. The discovery of the quantum, which is often called W^+ and W^- these days (though I adhere to my original name for it), procured a Nobel Prize for Carlo Rubbia and Simon van der Meer.

The electromagnetic and weak forces can be thought of as "flavor forces," since the electric charge varies with flavor and the weak force involves the changing of flavor. During the 1950's and 1960's a sort of quantum flavor dynamics was formulated, including quantum electrodynamics and a theory of the weak force. Among the successful predictions of quan-

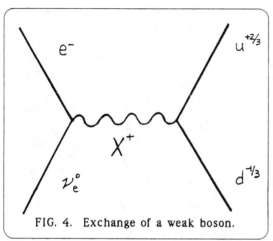

FIG. 4. Exchange of a weak boson.

tum flavor dynamics was that of a new flavor force that permits simple scattering processes for neutrinos, like

$$v_e^0 + p \rightarrow v_e^0 + p$$

and

$$v_e^0 + n \rightarrow v_e^0 + n,$$

or, more basically in terms of quarks:

$$v_e^0 + u^{+2/3} \rightarrow v_e^0 + u^{+2/3}$$

and

$$v_e^0 + d^{-1/3} \rightarrow v_e^0 + d^{-1/3}.$$

Here what is involved is the virtual exchange of a heavy, electrically neutral quantum called Z^0, as shown in Fig. 5. That quantum too, was discovered by Rubbia, van der Meer, *et al.*

Quantum flavor dynamics with four quanta (photon, X^+, X^-, Z^0) has recently scored considerable success in predicting the results of experiments, but from the theoretical point of view, certain aspects of the theory make it seem provisional and likely to be embedded in a bigger and better theory.

Let us summarize, in Fig. 6, what we have described so far: the two flavors e^- and v_e^0; the two flavors of tri-colored quark, $d^{-1/3}$ and $u^{+2/3}$; the flavor forces, including electromagnetism, acting on the flavor variable, carried by the photon and by the quanta X^+, X^-, and Z^0, and described by quantum flavor dynamics; and the color forces acting on the color variable (not present for e^- and v_e^0), carried by the colorful gluons, and described by quantum color dynamics.

Amazingly, Nature does not find this list of elementary particles to be sufficiently long. There are at least two ways in which we must extend it.

First, let us describe the electron and the neutrino, treated as two flavors, and the tri-colored **d** and **u** quarks, with their two flavors, as sort of family of particles. At higher masses, there are two more such families. (See Fig. 7.) The so-called muon μ^-, a sort of heavy electron discovered at

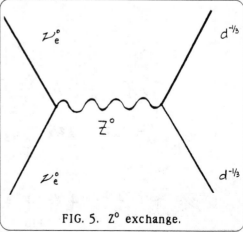

FIG. 5. Z^0 exchange.

Caltech by Carl Anderson and Seth Neddermeyer in 1937, turns out to have its own neutrino ν_μ^0, and there are two more flavors of quark to go with them, called the strange quark $s^{-1/3}$ and the charmed quark $c^{+2/3}$. (Note the frequently seen bumper sticker: PHYSICISTS DO IT WITH CHARM.) The tauon, τ^-, a still heavier electron, discovered at SLAC (the Stanford Linear Accelerator Center) around 1974 by Martin Perl *et al.*, appears to have its own neutrino as well, ν_τ^0, and seems to be accompanied by two more flavors of quark, $b^{-1/3}$ and $t^{+2/3}$. Actually, the existence of $t^{+2/3}$ has not been demonstrated experimentally, but it is certainly expected by theorists, who ought to commit *seppuku* if it is not found. (My col-

FIG. 6.

league "Murph" Goldberger, former President of Caltech, used to describe this process as " falling on our fountain pens." These days, fountain pens are scarce. Besides, we know that the ancient Roman hero who wanted to kill himself after a defeat had a trusty retainer hold his sword - could a graduate student hold the fountain pen steady enough?)

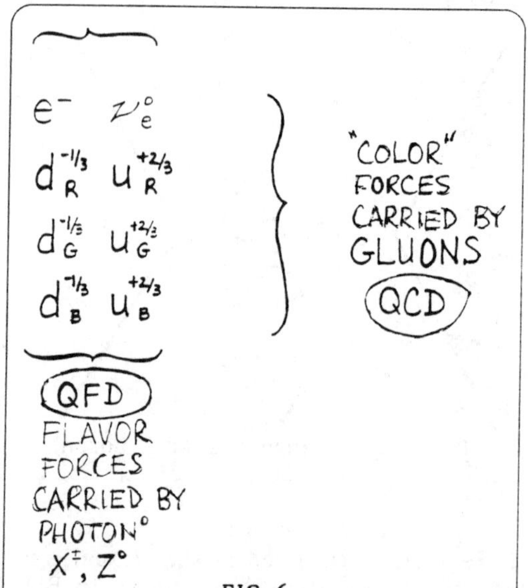

If we succeed in understanding the existence of the family made up of e^-, ν^0, **u**, and **d**, we must also explain the tripling of the family - who ordered

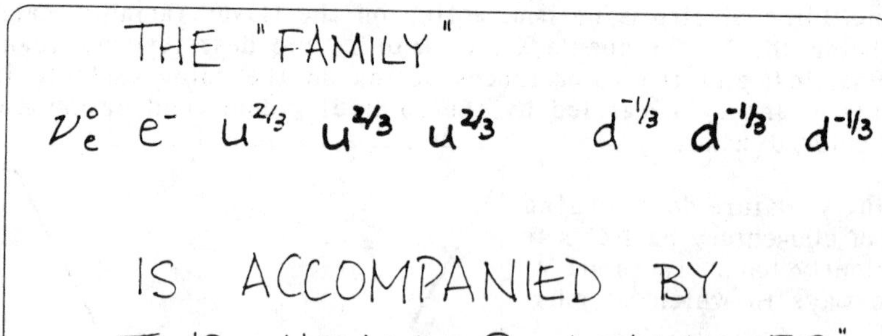

FIG. 7.

all those extra flavors?

Another way in which the list of elementary particles must be extended is by adding anti-particles. In quantum field theory there is an anti-particle with the same mass and equal and opposite values of electric charge and of certain other quantities. For some electrically neutral quanta, such as the photon and Z^0, the anti-particle is the same as the particle. For the other quanta we have listed, the anti-particles are already on the list; for example X^+ and X^- are anti-particles of each other, and so are, for example, red-blue and blue-red gluons. Thus, for the quanta, including anti-particles introduces nothing new. However, for the three families we have discussed, that is not the case - the anti-particles are all distinct from the particles, as indicated below using the first family as an example:

$$e^- \longleftrightarrow e^+$$

$$\nu_e^{\,0} \longleftrightarrow \bar{\nu}_e^{\,0}$$

$$u^{+2/3} \longleftrightarrow \bar{u}^{-2/3}$$

$$d^{-1/3} \longleftrightarrow \bar{d}^{+1/3}$$

We have now described $2 \times 3 \times 8 = 48$ members of the three families, including the anti-particles, plus 4 electro-weak quanta and 8 gluons, making 60 different particles so far. The layman will naturally ask, "Why should there be so many different kinds of elementary particle?" The elementary particle specialist asks exactly the same question.

One possible answer is that quarks, electrons, and so forth, and perhaps the quanta as well, are composite in their turn. Although it is true that they have not shown any sign of compositeness so far, the evidence for it could still turn up at very high energies. However, there is no convincing composite scheme that reduces drastically the number of fundamental objects needed.

Chairman Mao expressed an opinion on this subject, with which I suppose it was advisable to agree if you lived in China before the suppression of the "gang of four." He said that there would be an infinite regression, with nucleons composed of quarks, quarks of sub-quarks, and so on. No appealing theory of this type has turned up, and even in China, under the milder rule of the present leadership, I don't think the "infinite series of layers" is being actively pursued.

There is another approach, which need not involve seeking objects more fundamental than today's elementary particles. Instead, it uses the assumption that nature has chosen a special elegant mathematical scheme that involves a particular list of elementary particles, presumably the ones we know plus others. (Actually, it is not excluded here that the really elementary particles might be constituents of the ones with which we deal today, but that complication is probably unnecessary and I shall not discuss it further.)

The elegant mathematical scheme would involve a master equation for a single field with many components corresponding to the various elementary particles. In this approach, simplicity then lies not in economy of particles but in economy of principles.

The search for a single beautiful master equation evidently involves the attempt to unify the various forces. Unlike the material I have presented so far, what we are now discussing is highly speculative, not supported by a rich body of experimental evidence. It is natural to inquire how such speculation is carried out. The accompanying cartoon (Fig. 8.) shows the master at work, trying out various theories and discarding them until he finds the right one!

In fact, the most important tool of theoretician's's trade is not the foun-

tain pen, or the ball-point pen and mechanical pencil that have succeeded it, but the wastebasket. We write out the equations of promising theories on sheets of paper and then crumple up the sheets and throw them away when the theories turn out to be either self-inconsistent or else incompatible with some well-organized body of theory supported by unshakeable experimental evidence. Rarely, a proposed theory survives this process and goes on to be studied seriously and eventually tested experimentally.

During the 1970's, most speculation about the unification of the forces of Nature involved schemes that embraced quantum flavor dynamics and quantum color dynamics, together with new forces

FIG. 8. The master at work.
© 1977 by Sidney Harris -
American Scientist Magazine.

mediated by new quanta, in what we might call a "quantum unified dynamics." This task was not quite so daunting as it might appear, because both flavor dynamics and color dynamics belong to the same class of theories, called Yang-Mills theories after the two theorists who proposed them in the 1950's. What was being sought was a unified Yang-Mills theory, referred to by some theorists as a "grand unified theory" or GUT. (The name GUT is inappropriate, in my opinion, for two reasons: First, there was no previous unification, since quantum flavor dynamics exhibits only a mixing, not a true unification, of the electromagnetic and weak forces; and second, one should not speak of "grand unification" when gravitation is being ignored, because it doesn't include gravity, which is the most interesting force of all.)

Several related unified Yang-Mills theories look promising, but they are difficult to test because the effective unification takes place at an energy at least about 10^{13} (ten trillion) times higher than that accessible at today's largest accelerators. The theories thus represent a bold extrapolation from what is known, and we cannot hope to explore directly the energy range involved.

Fortunately, there are some predictions of unified Yang-Mills theory that can be tested at low energies, even though they depend on the properties of the theory at the enormously high energy of effective unification. The most important prediction of a typical unified Yang-Mills theory is that of proton decay.

Since the proton was known to decay, if at all, at a rate slower than 10^{-30} per year(!), most theorists had assumed that it was perfectly stable, although we had never found a good reason for its stability. In the early 1970's, with the advent of speculation about unified Yang-Mills theories, which predict rates of decay something like 10^{-32} per year, the rush to find proton decay in the laboratory was on. The search must be carried far underground, where false signals from cosmic rays are not a problem. Experiments are under way in a salt mine in Ohio, a silver mine in Utah, and a tunnel under the Alps, but so far no convincing cases of proton decay have turned up, even though the upper limit on the rate has been brought down near 10^{-32}.

Still, most theorists are betting on proton decay, perhaps at a somewhat lower rate, because of a point made long ago, even before unified Yang-Mills theories were proposed, by Andrei Sakharov, the Soviet physicist, weapons designer, and activist for peace and human rights. He pointed out that proton instability could help to explain the apparent preponderance of matter over anti-matter in the universe.

Meanwhile, a number of theorists have gone beyond thinking about unified Yang-Mills theories to working on the ultimate problem of elementary particle theory, the attempt to unify quantum flavor dynamics and quantum color dynamics (perhaps brought together in a unified Yang-Mills theory) with a quantum version of Einstein's "general relativity" theory of gravitation. Here we come, finally to the attempt to achieve a complete synthesis, to find a single theory that would describe all the forces and all the elementary particles of Nature.

It was Einstein who made unified field theory a very famous goal and tried hard to reach it. Even though he was the greatest theoretical physicist since Newton, he failed to find what he was seeking. He worked only with gravitation and electromagnetism, since in his time the weak and nuclear forces were poorly understood; although the electron was well known, he did not use it in his work, hoping that it would emerge as a prediction; he knew nothing, of course of quarks and gluons and other recently discovered elementary particles; and, most important, he ignored quantum mechanics, which he found philosophically objectionable. It is hardly surprising that Einstein did not succeed in this endeavor that occupied the last two decades or so of his life.

Today, the prospects look much brighter, and, in fact, in superstring theory, we have a promising candidate for a true unified theory.

Any correct theory must include Einstein's general-relativistic theory of gravitation as an approximation, the gravitational force must be carried by a quantum, which we call the graviton, and many properties of the graviton can be readily deduced from Einstein's theory. (For example, it is electrically neutral, it travels with the speed of light like the photon, and it has two units of a certain quantity called "spin.").

The graviton is extraordinarily difficult to detect, because the gravitational force is a very feeble between elementary particles, although most of us don't think of gravity as feeble while we are in the pull of a whole planet. (Even classical gravitational waves have not been directly detected so far, although there are experimental physicists working hard on that project.) However, the theoretical arguments for the graviton are overwhelming.

What a completely unified quantum field theory must do is to generalize a quantum version of Einsteinian gravitation, based on the graviton, to make it consistent, and to have it include all the other forces and particles.

We are now at the point in this article where we can begin to describe superstring theory, which has given us the only serious candidate for such a unified theory.

Superstring theory obeys a fascinating principle of self- consistency that is called the "bootstrap" principle because it allows the whole system of elementary particles, in a metaphorical sense, to "pull itself up by its own bootstraps." This principle imposes a very stringent condition on the theory. It was first proposed twenty-five years ago by Geoffrey Chew and Steven Frautchi in connection with theories of the nuclear particles alone.

I remember objecting, in the middle sixties, to one aspect of the bootstrap research of Chew and his collaborators, namely that they utilized only a very small number of particle types in their theoretical models. I suggested that it would be much better to try to include an infinite set of particles. Three Caltech postdoctora l fellows (Richard Dolen, David Horn, and Christoph Schmid) wrote a seminal paper on "duality" embodying this idea. Shortly afterwards the theorist Gabriele Veneziano produced an ingenious model of an infinite set of particles satisfying the "duality" version of the bootstrap conditions.

Unfortunately the Veneziano model had some apparently unacceptable features, but in 1971 John Schwarz and Andre Neveu, working at Princeton University and making use of some work of Pierre Ramond, developed the first version of superstring theory, which was to overcome the difficulties of the Veneziano model.

Thus work on superstring theory began fifteen years ago as an attempt to describe only the nuclear particles (neutron, proton, meson, etc.) which are now known to be composed of quarks, anti-quarks, and gluons. Even then many of us believed in a theory based on these constituents, but quantum color dynamics had not been fully developed, and there was a good deal of discussion of alternative theories.

At that time, the energy scale of superstring theory was thought to be around 1 GeV, near the rest energy of the neutron or proton.

As a theory of the nuclear particles, superstring theory was not notably successful. In particular, a certain theoretical particle kept cropping up, with special properties, which had no counterpart among the the real particles. All efforts to suppress this unwanted particle in the theory were unsuccessful.

Soon afterwards, around 1972, the present formulation of quantum color dynamics was achieved, and fairly soon it was acknowledged to be, most probably, the correct theory of the nuclear particles. Most theorists stopped working on superstrings. But John Schwarz continued to pursue research on the subject. In 1972, we invited him to Cal Tech.

Two years later, we had a brilliant young visitor from France, Joel Scherk, who collaborated with Schwarz on a radical reinterpretation of superstring theory. They looked again at the unwanted particle in the theory - electrically neutral, traveling with the velocity of light, possessing two units of "spin" - and decided it was not a nuclear particle at all - it was the graviton!

Studying superstring more closely, they found it contained a quantum version of Einsteinian gravitation as an approximation, along with many other particles and forces. They reached the startling conclusion that they were dealing with a possible unified theory of all the elementary particles and all the forces of Nature.

In order to effect the change from a (wrong) theory of the nuclear particles alone to a (possible) theory of all the particles and all the forces, Scherk and Schwarz had to alter the mass scale of the theory by an impressive factor, changing it from around 1 GeV to around 10^{19}, the fundamental energy scale of quantum gravitation. This "Planck energy" (around the rest energy of a postage stamp) is constructed from the fundamental constants of Nature: the velocity of light, the constant of quantum mechanics h, and Newton's constant of gravitation, G or κ. Corresponding to the "Planck energy" is the fundamental "Planck length" of about 10^{-33} cm. Just as the energy scale is about 10^{19} times larger than the rest energy of a neutron or proton, so the length scale is about 10^{19} times smaller than the size of a neutron or proton. In superstring theory, because it generalizes quantum gravitation, we are making even bolder extrapolation from attainable energies and distances than in unified Yang-Mills theory.

The word "string" appears in the name of the theory because the particles are described, on a length scale of about 10^{-33} cm, not as mathematical points (as in usual quantum field theory) but as tiny "strings," typically loops. (It was Yoichiro Nambu of the University of Chicago and his collaborators who first emphasized the string interpretation of the theory around the beginning of the 1970's.)

Analyzing such a string into modes, like a violin or piano string with its harmonics, we get the equivalent of an infinite number of kinds of point particle. Thus, in superstring theory, the number of elementary particles is infinite.

The infinite set of particles is described, however, by a single "string field" obeying a single fundamental equation.

We begin to understand why we find a large number of elementary particles. But why don't we find an infinite number? In the theory only a finite number of the particles have masses that are very small compared to the huge "Planck mass." These low-mass particles are the only ones that we have a chance of detecting.

The known particles, as we have seen, number about sixty (or sixty-one, including the graviton). Allowing for undiscovered ones, we may guess that the low-mass particles may number in the hundreds. Why should that be so? We shall have a hint of that. But why are the masses of the known ones so tiny (10^{-17} to 10^{-22}) compared to the Planck mass? That we do not know. Somehow such tiny numbers must be generated within the theory if it is to be correct, because superstring theory has "no free parameters," nothing to adjust to get agreement with observation.

The name "superstring" contains the prefix "super" because the theory possesses a broken symmetry called "supersymmetry" between the two great classes of elementary particles, "fermions" and "bosons."

"Fermions," named after Enrico Fermi, include the three families of particles that contain the electron, the quarks, and so forth. They obey the "exclusion principle," which means they cannot stand to be in the same quantum state at the same time.

"Bosons," named after the Indian (Bengali) physicist S.N. Bose, include the quanta such as the photon, the gluons, and the graviton. They *love* to be in the same state at the same time.

The fact that photons are bosons (pointed out almost seventy years ago by Bose and by Einstein) supplies the basic principle of the laser beam,in which all the photons have, to a high accuracy, the same energy and the same direction.

The broken "supersymmetry" between fermions and bosons means that in accelerator experiments at sufficiently high energies (and we hope that these high energies will be low enough to be attainable in experiments at accelerators) we must find partners as follows:

for the photon (a boson): a heavy " *photino*" (a fermion);

for the gluons (bosons): heavy "*gluinos*" (fermions);

for the electron (a fermion): a heavy *selectron* (boson):

for the quarks (fermions): heavy *squarks* (bosons): etc.

Please do not blame me for names like "squark."

These "superpartners" of known particles are being sought at existing accelerators (Fermilab near Chicago and the SPS at CERN, the European Council for Nuclear Research, near Geneva, Switzerland) and will be sought at higher ener- gy accelerators (LEP, being built by CERN in France near Geneva, and the SSC or super- conducting super-collider, which we hope the U.S.A. will build).

A remarkable property of superstring theory is that its self- consistency seems to fix the number of spatial dimen- sions, and that number is not three. Instead, the theory ap- pears to work only in nine spatial dimensions, giving a ten-dimensional space-time. Joan Cartier, a theoretical physicist who is also a cartoon- ist, has shown, in the accompa- nying drawing, how a lecture on this subject appears to her. (Fig. 9.)

© *J. F. Cartier*

FIG. 9. "AT THIS POINT WE NOTICE THAT THIS EQUATION IS BEAUTIFULLY SIMPLIFIED IF WE ASSUME THAT SPACE-TIME HAS 92 DIMENSIONS".

What has become of the other six dimensions? If they are taken literally, it must be supposed that, unlike the familiar dimensions, they have not out- grown the size they had when the universe was in the first tiny fraction of a second of its expansion. Instead, at every point in three-dimensional space, the extra dimensions are curled up into a minute structure that has roughly the size of the Planck length, around 10^{-33}cm. This is so small that the extra dimensions do not play much of a role as such: but it turns out that the nature of the six-dimensional structure is crucial in determining the broken symmetries of the system of elementary particles. Much impor- tant research on that subject has been carried out by Edward Witten and his collaborators at Princeton University.

To get some idea of how to think of extra spatial dimensions, imagine that we are all "flat-landers" with only two dimensions, living in a two-dimen- sional world, and that some flat savant announces to us that he has good news and bad news: the good news is that we all have a new dimension, height, that we never knew about before: the bad news is that no flat-land- er and no place in flat-land has a height greater than 10^{-33} cm.

Since 1986, the notion has gained ground that in the solution to the superstring equations the extra dimensions may not have to be taken literally; perhaps they can be treated as mathematical constructs instead. The behavoir of the extra dimensions remains, however, intimately connected with the symmetry pattern of the elementary particles and their interactions.

During the period from 1974 to 1983, there was considerable activity in the world of theoretical physics on supersymmetry and related questions, but very little on superstrings outside Caltech. We were continuing to maintain nature reserve for an endangered species, the superstring theorist, with the help from the Department of Energy and from the Fleischmann Foundation. That Foundation gave us generous support just as it was liquidating itself, a convenient arrangement since we had only to account for the legitimacy of our expenditures and not for the relevance of our speculations.

Starting around 1980, a frequent visitor was Michael Green, from London. He and Schwarz have been collaborating for the last seven years or so, and in 1983 and 1984 they produced some exciting results (some of them achieved during summer work in the stimulating atmosphere of the Aspen Center for Physics, Aspen, Colorado.)

To appreciate the first one, it is necessary to understand that quantum field theory is usually plagued by the occurrence of infinite quantities in calculations. Naive calculations typically take the form:

FINITE TERM + INFINITE CORRECTION + WORSE INFINITE CORRECTION + ...

Years ago, I had become so used to working with such formulae that someone had asked me "Murray, what would you do if a calculation converged on you?"

Now in successful theories like quantum color dynamics or quantum flavor dynamics, the infinities can be absorbed into a few quantities, which can be arbitrarily set equal to finite, but calculable values. We obtain the finiteness at the price of the uncalculability. This process, which is called "renormalization," amounts to sweeping infinities under the rug, as Joan Cartier shows us in another one of her cartoons. (Figure 10.)

Now a straightforward quantum version of Einstein's general- relativistic theory of gravitation leads to infinities that are not even renormalizable. This has been known for many years for the case where the theory treats not only the graviton but other kinds of matter as well. It has recently been shown by Marc Goroff and Augusto Sagnotti that even for gravitons alone there are unrenormalizable infinities. In other words, no Band-Aid will fix the quantum version of Einsteinian gravitation. A radical generalization is needed, such as superstring theory provides.

Schwarz and Green showed, in 1983 and 1984, that superstring theory seems to give only finite results. No infinities appear in the calculations,

FIG. 10. "ALL RIGHT, RUTH, I'VE GOT THIS ONE ABOUT RENORMALIZED." ©J. F. Cartier.

not even renormalizable ones. Moreover, superstring theory is the only known generalization of quantum gravitation that is free of unrenormalizable infinities.

In 1984, Schwarz and Green made another discovery. Superstring theory was constructed using a system of elementary particle symmetries called a "Lie algebra" after the nineteenth century Norwegian mathematician Sophus Lie. These symmetries are present in addition to supersymmetry. It had been thought that there was an infinite variety of such algebras that could be used, for example, one with three independent symmetries, or one with six, or one with ten, and so forth. But Schwarz and Green showed that self-consistency restricted the choice to just two possibilities, called SO_{32} and $E_8 \times E_8$, each with 496 independent symmetries!

The discovery that the choice of symmetry system was restricted to only two, both with 496 symmetries, created great excitement in the world of theoretical physics, especially coming just after superstring theory was known to be finite.

We may well ask, "How could a theory based on the number 496 be wrong?"

In any case, theoretical physicists around the world rushed to jump on the superstring bandwagon. In December 1984, a team of four theorists at

Princeton University (David Gross, Jeffery Harvey, Emil Martinec, and Ryan Rohm, known, of course as the "Princeton string quartet") found a new form of superstring theory, using the number 496, and based on the symmetry system $E_8 \times E_8$. Their version, which they call the "$E_8 \times E_8$ heterotic" superstring theory, looks like the best one for understanding the structure of the elementary particle system as revealed so far by experiment. Why has Nature should have chosen this form of superstring theory rather than the two others that also utilize the number 496, we do not know. That is a puzzle that we shall have to solve.

Let us summarize what we have discovered about the virtues of superstring theory. It gives us:

- an elegant, self-consistent quantum field theory;

- generalizing Einstein's general-relativistic theory of gravitation treated quantum-mechanically;

- in the only known way that does not produce infinities;

- parameter free;

- based on a single string field;

- but yielding an infinite number of elementary particles;

- some hundreds of which would have low mass (although we don't know why they would be so very low!);

- with the underlying symmetry system essentially determined;

- and with the symmetry breaking connected with the behavior of some extra, but perhaps formal, dimensions.

There are certainly some indications that our colleagues may have found the "Holy Grail" of fundamental physics.

But only calculations and their comparison with experiment will allow us to tell. In our of science, no amount of eloquence will save a wrong theory.
□□

MAURICE GOLDHABER:

Once a student stopped me at a meeting and asked me, "Do you believe squarks will ever be found?", and I said, "Only Sgod knows". We now have a break of 20 minutes, and we want you to get back punctually because we have to meet the deadline of 5 PM. (Break taken.)   We now start the last session of this meeting. This has been a year of super physics, and the next two talks will deal with super physics which was largely developed this year. It is appropriate that the man who discovered a star -- a man who has the distinction of having discovered a star which does not exist anymore, Nicholas Sanduleak, is in the astronomy department here at Case Western Reserve University. He has the distinct honor of having a star named after him which isn't there any more. Now you can all perhaps go home and say I also have a star named after me which doesn't exist.

Our next speaker is Robert Kirshner, a professor of astronomy at Harvard. He received his doctorate at the California Institute of Technology, then was for many years at the University of Michigan, and he is now at the Harvard-Smithsonian Center for Astrophysics. He has been long interested in very large objects in space and in supernova phenomenon. He recently made important investigations of the radiations from the supernova emitted into ultraviolet. Professor Kirshner.

SN1987A: The Supernova of a Lifetime

Robert Kirshner

HARVARD-SMITHSONIAN CENTER FOR ASTROPHYSICS

Cambridge, Massachusetts 02138

I want to describe observations of the supernova that we call 1987A, or, since it was discovered by a Canadian, *1987, eh?* This is the brightest supernova that's been seen in 383 years. So the title that I chose for this talk, *The Supernova of a Lifetime*, is quantitatively inappropriate; it's the *Supernova of Seven Lifetimes*. The last one that was visible to the naked eye was seen by the nude retina of Johannes Kepler in 1604. He did not discover the supernova, but he wrote the best papers on it, and many of us are keeping this in mind.

The supernova occurred in a nearby galaxy, the Large Magellanic Cloud (LMC), a paltry 170 000 light years away. The LMC is named after Ferdinand Magellan, or possibly *by* Ferdinand Magellan, who saw it on his famous 1521 world tour, once he got to Southern latitudes. It is a satellite of our own Milky Way Galaxy. Knowing the distance to the supernova is a great advantage, since it allows observed quantities to be connected to the properties at the source.

Since Kepler's time there have been some new techniques, in addition to the eyeball, which allow us to analyze the light from stars, and which allow us to enrich our picture of what exploding stars really are. Although it seems gratuitous to talk about telescopes for a naked eye supernova, nevertheless, we have been using large and small telescopes (often with dark neutral density filters to cut down the light) to measure the optical light from the supernova. It has also been measured with radio telescopes, with detectors on a geosynchronous satellite (not available to Kepler), and even with such bizarre things as tanks of water deep in salt mines. All these techniques provide tests for a well developed picture that unites many astrophysical themes.

A supernova results from a massive star at the end of its life. In the case of 1987A we know the name of the star that exploded and we have good evidence that it was a massive star. The center of the star is thought to collapse, so the source of energy for the supernova explosion is gravitation. It emits most of the binding energy as neutrinos and forms a neutron star. The central part of the supernova explosion is really not an explosion, but an implosion. This also has been tested in a unique way for SN 1987A through neutrino observations. While all that horrible business is going on inside, a little energy is coupled to the outside of the star which ejects the outside of star and heats it, so that it shines with 10^8 solar luminosities and blasts off its envelope at 1/10 the speed of light. This great luminosity allows us to study supernovae at cosmologically interesting distances and use them to measure the geometry of the universe, but here we're concerned with understanding the details of a nearby example.

The heavy elements are created both in the course of energy generation for stars and also in the explosion itself, where there is synthesis of
heavier elements from iron seeds. Supernovae have the additional very
valuable property of destroying themselves and disgorging their innards
into the interstellar gas. Supernovae that exploded over five billion years
ago synthesized the material of our own galaxy, of our own sun, of our own
solar system, and indeed of the ink in this book. (To say nothing of the pin-
nacle of evolution, the reader!) We have inherited a rich legacy of heavy
elements from the generations of stars which lived and died before the sun
was formed. What's more, the supernova explosion deposits of a lot of kinetic and thermal energy in a small place in the interstellar gas and drives a
strong shock into the surrounding clouds. Those shocks are important in

FIG. 1. The Las Campanas Observatory.

triggering additional star formation, so that this entire cycle: the origin of
stars, the death of stars, and the chemical enrichment of the universe, is
one in which supernovae play a central role. Supernova 1987A gives us a
chance to test this picture.

We've heard a lot of thinly veiled humor about trips to Stockholm.
Astronomers do not make trips to Stockholm, they make trips to Chile,
which provides a refreshing change of seasons, a wonderful astronomical
climate and access to Southern skies. Figure 1 is the scene on the fringe of
the Atacama Desert where the supernova was discovered, not by the 100"
telescope of the Las Campanas observatory (which is why it's not shown),
not at the 40" telescope at the left of the Las Campanas Observatory, nor
even at the 24" telescope of the University of Toronto. In this picture you
can see the little dog house, which contains the telescope where the supernova was discovered. Shown is the stone " *Casa Canadiensis*", where the in-

mate dwells who keeps the Toronto station going. There are very few distractions on Las Campanas. As Ian Shelton, the resident observer, remarked, there is a woman behind every tree on Las Campanas.

Ian Shelton used a small and antique piece of equipment in the heroic mode, that is, by standing outside in the cold. It's a camera that was put down there when the Carnegie Institution developed the Las Campanas site. Someone had enough confidence in Newtonian physics to gamble that Halley's Comet was going to come back in 1986, and they thoughtfully installed a small wide field camera.

Ian Shelton had taken many photographs of the comet and decided that he had better get some calibration plates by photographic the LMC. On one of his plates, he noticed a bright star which did not belong, as shown in Fig. 2. The Large Magellanic Cloud is an irregular galaxy, not a beautiful spiral like our own, but there are those who love it. It has very large and spotty regions of star formation, the most conspicuous of which is the subject of this portrait. The real connoisseur looking at this picture would say, "*Mon Dieu!*, there's is a star here doesn't belong". This star, the 1987A supernova, was picked out by Shelton without having a comparison photograph in his hand. The figure also shows the extraordinary environment of the supernova. This is the Tarantula Nebula.

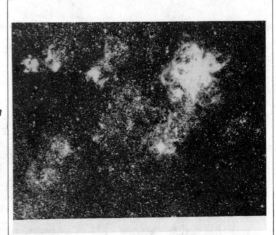

FIG. 2. Tarantula Nebula and SN 1987A (European Southern Observatory photograph.)

The Tarantula nebula is a large region where hydrogen gas is ionized by ultraviolet light from young, massive stars which presumably formed out of this gas cloud. It is a nursery for stars. It is also a stellar graveyard because the massive stars live fast and die young in the place where they're born. There's every reason to think (now that we've begun to autopsy the stellar corpse) that SN 1987A resulted from one of the generation that's been formed in this neighborhood, lived out a very brief life of about 10 million years and come to a notorious end. Our own lives are so short: 10^{-5} of this stellar span, that it is rare indeed when we get to see the death of a nearby massive star.

The Large Magellanic Cloud has been studied extremely well. Armed with accurate measurements of the supernova position, we can go back to old data and see what was there before the supernova exploded. When you pull out the old pictures of the Large Magellanic Cloud there is an arrow on

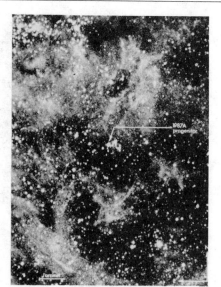

FIG. 3. 1987A Progenitor. (Southern Observatory photograph.)

them which points to a rather conspicuous star, and it's labeled "1987A PROGENITOR" (Fig. 3.).

What makes a massive star explode? One idea, which has been floating around for over 50 years is the gravitational collapse to a neutron star. Figure 4 is from the Los Angeles Times of January 19, 1934. In one dense paragraph we have the most compact prediction in the history of science, according to Fritz Zwicky, Swiss physicist. He says, "exploding stars ...burn with a fire equal to one hundred million suns" (this is the good part) "and then shrivel from half million mile diameters to little spheres, 14 miles thick." Caltech physicist Fritz Zwicky and his partner, Mount Wilson Observatory astronomer Walter Baade were postulating gravitational collapse to neutron stars in 1933, *just after the discovery of the neutron.*

Figure 5 illustrates one of the main points made by Hans Bethe: that stellar evolution and supernova explosions are the true alchemy. Stars achieve the ancient alchemist's nightmare: turning gold into lead, while also synthesizing other elements. The products of hydrogen burning for the exploded star can be detected in the debris, as well as the more exotic things that only a supernova can make.

What does the supernova become later? We know from looking at the sites of remnants of supernovae: Fig. 6 shows a supernova remnant in the LMC excitingly named 0540-69. It has oxygen-rich debris from deep inside a massive star expanding at a few thousand kilometers per second, and a spinning neutron star within flashing 20 times a second. It was

Fig. 4. " Be scientific with Ol' Doc Dabble".

probably the previous supernova in the LMC, perhaps 1000 years ago.

The supernova story has the big science of the best equipped observatories and particle detectors, and the small science of amateurs and old, small telescopes. Some details of the discovery of the supernova have important implications for the neutrino observations. Shelton's discovery took place on the 24th. The supernova was also discovered independently in New Zealand, by Albert Jones, a very avid amateur astronomer. Curiously, there were photographs of the supernova taken in Australia the day before the discovery. When inspected on the 24th they showed that the supernova had already reached sixth magnitude on the 23rd. That's about the dimmest star that a really acute observer can see from a dark site. Now Mr. Jones, who discovered the supernova independently on the 24th, took a look at the LMC on the 23rd and did not see the supernova. This nonsighting is as important as Sherlock Holmes's dog that did not bark: it took place just before the Australian photograph, so we can pin down when the supernova brightened. The supernova brightened after the neutrinos (presumably generated in the core collapse) reached Earth. There were two coincident neutrino events, the Kamioka device in Japan and the IMB detector in Fairport, Ohio. Each saw a burst of neutrinos about 3 hours before the supernova got bright. This interval is the time it takes for a shock wave to reach the surface of the star, heat it and make it start to expand.

FIG. 5. "The True Alchemy". [S. Woosley, U.C. Santa Cruz.]

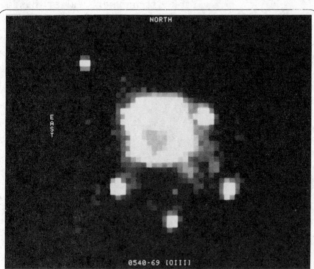

FIG. 6. Supernova remnant 0540-69 in the LMC. [Cerro Tololo Inter-American Observatory.]

It was wonderful to have a believable neutrino detection, as Prof. Reines showed, and even better to have agreement between the widely separated experiments. However, there is an embarrassment of riches: a surplus neutrino detection. An experiment in the Mont Blanc tunnel is operated specifically to look for cosmic neutrinos. The group that runs it reported that they saw 5 events in their machine, 4.6 hours earlier than the events that were seen in the Kamioka and IMB devices. The evidence from the optical light curve, when combined with a model for the star that exploded, makes it very unlikely that the Mont Blanc event deposited any energy in the star; otherwise it would have brightened hours earlier, and that would have been seen by Bernard Jones. So the amateur observations help constrain the fanciest results from the big science detectors.

Those events took place on February 24th 1987, and the word was passed to the Central Bureau for Astronomical Telegrams, located in the Center for Astrophysics at 60 Garden Street in Cambridge and run by Brian Marsden. No tree falls in the astronomical forest until Brian hears it. On that Tuesday morning, I received a telephone call from Texas alerting me to the supernova of a lifetime -- I didn't find out from Brian Marsden, whose office is one hundred yards from mine. Those of you who have been to 60 Garden Street will understand: it is noneuclidian; there is no shortest distance between two points in that building. I got a telephone call from Craig Wheeler in Austin. He said, "Have you heard about the supernova in the Large Magellanic Cloud?" "Ha-ha-ha", I said, "you fooled me like this once before"; which he had at a meeting in Sicily. Craig and his unindicted co-conspirators had concocted a false report of a supernova, which lead me to change my travel plans, to try to rush home. I was not going to be fooled again. I went to check with the official grapevine.

I went through the twisting passageways to Brian Marsden's office with the utmost probity. As I walked in, he was talking on the telephone, typing on his keyboard, and the teletype was clunking away in the corner, so I knew that there was circumstantial evidence that some violent event had burst in on his Tuesday morning.

My own work involves observations of the supernova with the IUE (International Ultraviolet Explorer) satellite. I had written a proposal to NASA two years earlier, saying if there happens to be a supernova observable with IUE, I'd like to have a look, and if it happens to be especially bright perhaps more extensive observations would be warranted. I do this every year, like sending Christmas cards to distant relatives. They said, "Fine, you can be a Target of Opportunity, but we're not giving you any money until there's a supernova." That was fair enough, and opportunity knocked rather vigorously on February 24, when we began ultraviolet observations.

IUE is a geosynchronous satellite, which makes control easy. It is above the atmosphere to escape the pestiferous ozone that absorbs the UV. Of course, we're making progress in bringing ultraviolet astronomy down to Earth through atmospheric chemistry, but that's another story.

IUE is an ultraviolet satellite, but it also has an optical photometer to measure the brightness of an object. In Fig. 7 I show a plot of the magnitude (a

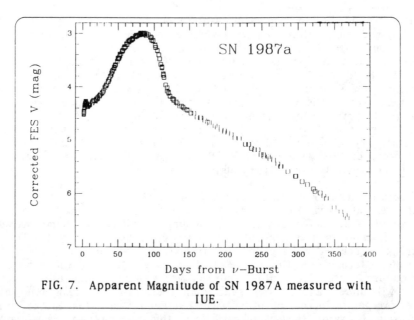

FIG. 7. Apparent Magnitude of SN 1987A measured with
IUE.

unit used by astronomers to keep the physicists uneasy-- actually 2.5 times
the log of the flux) which tracks the evolution of the supernova.   It
brightened rapidly in the first couple of days, paused, and then brightened
up to third magnitude, which is the brightness of the stars in the Big Dip-
per, three months later, about the 20th of May. The observed luminosity,
adding up UV, optical and infrared flux, corresponds to 10^8 solar luminosities. This is an impressive energy output, but only 10^{-4} of the energy released in the supernova explosion.

Supernovae that have hydrogen lines in their atmosphere, the kind we
call Type II, usually brighten to a maximum in the first 10 or 20 days, with a
peak luminosity 10 times brighter than SN87A. One of the initial mysteries
was why 87A was dimmer than others. Were we looking at a different phenomenon, or the same phenomenon in a different package? As we'll see,
it's the second possibility which looks more likely. The supernova faded in
June and then went into a long exponential decline, which is a clue to the
energy source for supernovae. Radioactive decay of elements synthesized
in the explosion is the likely explanation.

The supernova never got brighter in the ultraviolet. At the first moment
we saw it, it was declining due to adiabatic cooling of the expanding supernova atmosphere. When the shock from the center of the star hit the surface, calculations show it heated the surface of the star to roughly 200 000 K
but that temperature relaxes as the star expands with a timescale of an
hour. The observations in Fig. 8 show that the flux in the ultraviolet, on
the short wavelength side of the Planck curve for these temperatures, was
declining very precipitously: a thousand-fold in the first three days. The
flux in the UV declined, but did not reach zero. There was additional light in
our field of view coming from stars next to the supernova.

Figure 9 combines IUE data with visible data from Cerro Tololo Inter-American Observatory. It goes from about 1200 angstroms out past 9000 angstroms in the near infrared. You can see the overall energy distribution, and also see there is blue-shifted absorption and emission at the 6562 angstrom wavelength of H alpha. We know how lines like that get formed from absorption in a rapidly expanding atmosphere: it is strong evidence that this star really is blowing up as well as getting bright.

The velocities are remarkable. The absorption is shifted by 30,000 km/sec. The expanding atmosphere is moving out at one tenth the speed of light. So the star is truly blown apart. The kinetic energy is about 100 times the light output: of order 10^{51} erg, but this is small compared to the binding energy of a neutron star.

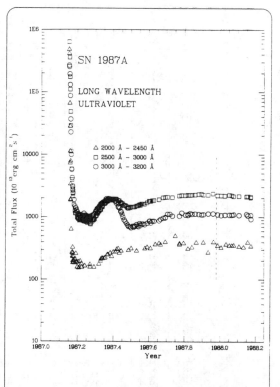

FIG. 8. Ultraviolet light curve for SN 1987A.

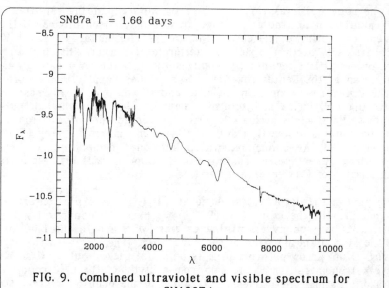

FIG. 9. Combined ultraviolet and visible spectrum for SN1987A.

TABLE III (cont.)

Star	Chart	R.A.	Minu. Dec.	Sp.	Mpg	Other designations
				-69° Zone (cont.)		
181	45b	$5^h 32^m 5$	69°14'	OB	12.0	
182	45b,53	33.2	24	F2I	11.5	269735, FDD
183	53	33.3	31	WR:	14.1	
184	53	33.7	20	OB	13.0	
185	53	33.8	08	OB:	12.7	
186	53	33.9	28	OB	12.6	
187	53	34.1	33	OB	12.5	
188	53	34.3	00	A0I	11.4	269762, FDD
189	53	34.3	04	OB	12.3	FDD
190	53	34.5	02	A7I	12.5	
191	53	$5^h 34^m 6$	69°46'	WR	13.4	37680, WS 35, L-286
192	53	34.7	11	OB	12.6	
193	53	34.8	48	OB	11.7	269769, -69°392, W 28-3
194	53	34.8	47	OB	12.2	W 28-10, L-289
195	53	35.1	15	OB	13.1	
196	53	35.1	39	OB	12.4	W 28-32
197	53	35.2	45	OB	12.2	269784, W 28-29
198	53	35.3	45	WR:	14.1	W 28-30
199	53	35.3	00	OB	12.8	
200	53	35.3	46	OB	11.3	269786, -69°399, W 28-34
201	53	$5^h 35^m 5$	69°42'	OB(e)	10.9	37836, R 123, W 28-39, S 124, L-291
202	53	35.6	18	OB	12.2	
203	53	35.6	15	OB	12.5	
204	53	35.6	28	OB	11.6	
205	53	35.7	43	OB	12.2	W 28-47, L-294?
206	53	35.8	08	OB	13.3	
207	53	35.9	12	WR	14.3	269818, WS 36
208	53	35.9	30	OB	12.8	
209	53	36.1	32	OB	11.7	
*209a	53	36.1	13	WR+OB	10.9	269828, -69°409, WS 38
210	53	36.2	03	OB	13.1	W 27-27
211	53	$5^h 36^m 3$	69°25'	OB	11.0	269832, -69°414, FDD
212	53	36.3	13	OB	12.0	
213	53	36.4	12	OB	12.5	
214	53	36.4	33	OB	12.3	
215	53	36.5	08	OB	11.7	269846, -69°417
216	53	36.7	24	Be	11.6	37974, R 126, S 127, AL-361, -60°402
216a	53	36.8	50	OB	13.1	
217	53	36.8	30	OB	12.1	
218	53	36.8	31	OB	12.0	
219	53	36.8	10	F2I:	12.0	269860, -69°422, FDD
220	53	36.9	31	Bep	11.4	269858, -69°427, R 127, S 128, AL-363, FDD
221	53	$5^h 37^m 0$	69°31'	OB	10.1	269859, -69°428, R 128
222	53	37.1	27	WR	13.4	38030, WS 39
*223	53	37.1	13	WR	14.2	38029, WS 40
224	53	37.1	13	OB	11.8	

FIG. 10. Part of Sanduleak's list of LMC stars.

Since the supernova was relatively nearby, and expanding rapidly, I did a quick calculation to predict its angular size. Michelson pioneered the measurement of stellar diameters, using an interferometer at Mount Wilson. Despite its early birth, the field of stellar interferometry has had a long infancy, and is still a technically young subject. Even so, I called our local pediatrician at the Center for Astrophysics, Cos Papaliolios, and told him that there was a supernova in the LMC and it would be of interest to measure its angular size. I also told him that my calculation showed it was impossible to do it right away. He wisely accepted the information and ignored the conclusion, setting out for Chile, and making a very interesting measurement of the supernova which showed a companion, the mystery spot, shining in very close proximity to the supernova itself. No satisfactory explanation of this observation exists.

One part of the dogma for supernovae was that the explosion erupts in a massive star. In the case of SN1987A, we have the opportunity to check whether that is true by examining catalogs of stars in the LMC to learn about the star at the position of the supernova. The star with the accusatory "progenitor" arrow pointing to it is on a list picked out by Case Western Reserve astronomer Nick Sanduleak from a study of the LMC conducted 20 years ago. It's just one of many LMC stars: in the band 69 degrees south of the equator it is number 202. In case you want to stick around for the next million years, the other 1273 in Sanduleak's catalog are all going to explode too. The table lists the position of the star 202, something about its brightness or magnitude, something about its spectrum, and, like a politician with something to hide, no comment in the "Comment" column. There was nothing to distinguish this star; nothing to point out that this was the next star to explode. (You might find that ominous!) The identification of Sanduleak -69 202 as the star that exploded marked the first time we have known the name of a star that exploded. More seriously, we know it was a massive star, about 20 times the mass of the sun, furiously burning away at 10^5 solar luminosities before the fatal day when its core collapsed. The local Cleveland headline, "Sanduleak explodes", really is true-- of the star but not of Nick.

Now let's have a look at the inside story. A star's biography is the story of nuclear binding. The sun cooks hydrogen into helium starting with the proton-proton reaction and releases the binding energy by forming more tightly constructed nuclei. For a more massive star, hydrogen fuses to helium through a reaction which involves carbon, nitrogen and oxygen, the CNO cycle. One effect of hydrogen burning by the CNO cycle in a massive star is to enrich the nitrogen at the expense of carbon and oxygen. We see hints this really happened for Sanduleak 69 202.

Massive stars use the ashes of hydrogen burning as the fuel for the successive stages of nuclear fusion: use the helium waste as a fuel to make carbon, use the carbon as fuel to make oxygen and neon, and so on all the way on up to iron. Each of these steps takes place at a higher temperature because the Coulomb barrier for these nuclei is higher. Each also takes place at a higher density. And most importantly, from the point of view of the outside of the star, the duration becomes very short. For example the central temperature for a star burning helium to carbon is about $2\ 10^8$ K and that burning stage lasts for about a million years, whereas hydrogen burning for a 20 solar mass star lasts for about 10 million years. The successive stages become very much shorter. From carbon burning to the end is only six hundred years. The outside of the star does not have time to respond to changes that take place inside. Once the star gets on this slippery slope, down to making iron in the inside, the outside of the star responds very little. That means the events in the center of the star are not reflected in whether the star is a red supergiant or a blue supergiant. The bomb that goes off in the center of the star is not closely coupled to the shape of the suitcase in which it's enclosed.

So here's the idea: carbon, oxygen, silicon and the other elements all the way up to iron are created by fusion. It costs energy to make the heavier elements, although supernova explosions have the energy and free neutrons to do that. That's why they're implicated in the synthesis of the heavy elements.

Catastrophe results because the center of the star has no more resources for nuclear energy once it's iron: the nuclear turnip from which no blood can be squeezed. The star loses energy through radiation from the surface. Even more important is the neutrino flux from the very high temperature region right in the core. The star has severe energy losses, and no energy sources. This is a recipe for disaster.

The iron core begins to contract and heat, but instead of igniting a new fuel, the iron photo-dissociates, the neutrino losses skyrocket, and a disaster ensues in which the center of the star collapses. The center of the star forms a neutron star, a little sphere 14 miles thick, when it is finally stopped by the strong force. Most of the 10^{53} ergs of binding energy of the neutron star is released in the form of neutrinos. At least that's the story that we've been telling ourselves for a long time. The first observational check has come in the case of SN 1987A. The nuclear cooking is supposed to take place in the material that doesn't quite fall into the core. It's heated to high temperatures as a shock goes out through the star. That material comes out, and we have a chance to look at the insides of a massive star. Last, but not least, the outer part of the star explodes. Some small fraction of the energy gets coupled into the star itself, and gets blasted off to make the bright and violent display we observe. Let's look at the energy budget. About 10^{53} ergs is thought to come out in the form of neutrinos, about 10^{51} ergs in the form of kinetic energy, and the light from the supernova is actually a fairly small fraction of the total: 10^{49} ergs in light.

Several things about this supernova were peculiar: it wasn't as bright as others, the color seemed funny, the velocities were unusually high. All of these are due to the fact that SN 1987A, unlike the hundred or so supernovae in other galaxies which we have observed reasonably well, did not disrupt a red supergiant, a cool star with a very extended atmosphere. The Sanduleak star, is known from observations to have been a blue star with a surface temperature of about 15 thousand degrees. Details of the light which were peculiar and puzzling at first turn out to be the result of just this one simple fact: the radius of the blue star was smaller than we normally see when the center collapses in a comparably luminous red star. Since this collapse in the center is fairly well uncoupled from the structure of the envelope, we're really learning about the same physical events in the center, whether we see the explosion of a red supergiant or a blue supergiant, but the external details are different. Nevertheless, an interesting question is why was Sanduleak -69 202 blue when it blew? Did it in fact have a more colorful past? We'll return to that point in a little while.

But first, let's consider core collapse. What are the predictions about core collapse? You release the gravitational binding energy of a neutron star (about 10^{53} ergs). You can calculate the radius at which the neutrinos would escape, and calculate a temperature for them; it's a few MeV (which is 10 000 000 times hotter than hell if brimstone melts at a few thousand K). You can calculate the diffusion time for the neutrinos. Although the neutrinos can go through a light year of lead, and have no trouble penetrating the earth, neutron stars are a lot denser than lead. After all, in lead, the nucleus occupies a tiny fraction of the volume for each atom, but in neutron star matter, the nucleons are cheek by jowl and the entire star is at nuclear density. The neutrino mean free path in a neutron star is about 3 centimeters. The neutrinos don't just stream out, they diffuse out in seconds, not the light travel time across a 16 km neutron star. In Ohio and in Japan, neutrinos were seen on the 23rd of February. Figure 11 is a record of the events. Although the number of events seen is small, a few numbers

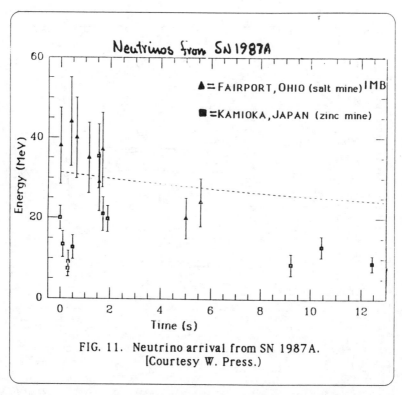

FIG. 11. Neutrino arrival from SN 1987A.
[Courtesy W. Press.)

can be extracted from the observations. First, the emitted energy at the source is about 10^{53} erg. You can see that the energy is up in the 10, 20, 30, 40 MeV range. This is the high energy tail of that 5 Mev thermal distribution. The threshold for the detectors are 10 MeV or more. The duration of the neutrino event is not a millisecond; it's on the order of several seconds. So the diffusion picture for neutrinos, which comes from the neutral current theory that we heard a little bit about today from Prof. Gell-Mann, is also confirmed. Finally, we note that the correlation between arrival time and neutrino energy is weak. If neutrinos had a large rest mass, the more energetic would get here faster. A conservative upper limit is that the neutrino rest mass must be less than about 25 eV.

Now let's return to the lurid past of Sanduleak's star. For example, the calculated evolution of a 20 solar mass star by Stan Woosley, of UC Santa Cruz, shows the path it takes in the H-R diagram before exploding. This is shown in Fig. 12. You can see that it spent some time over on the right as a red giant. One interesting sidelight is that red giants can lose mass from their atmospheres relatively easily. The material is at such a large radius that it is very nearly unbound anyway, and it doesn't take much for these stars to puff off their outer layers in a stellar wind. We want to investigate whether the Sanduleak star had an episode of mass loss. From ultraviolet emission lines (Fig. 13) we actually see slow moving material around the star which was excited to fluorescence by the initial ultraviolet pulse of the supernova explosion. That shell of gas emits most strongly in lines of nitro-

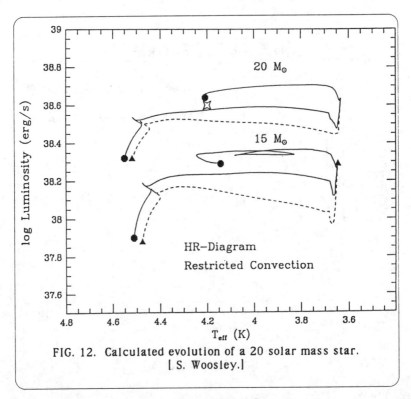

FIG. 12. Calculated evolution of a 20 solar mass star.
[S. Woosley.]

gen. This is significant because this is the nitrogen-rich material expected
as the result of CNO cycling in a red giant star. Quantitatively, N/C is about
10 here, and about 1/4 in the usual interstellar gas, so nitrogen is enriched
by a factor of 40. Since the lines are narrow, the material is moving less
than 30 km/sec. It is not inside the opaque atmosphere of the star, so we
think it's material which was lost by the star when it was a red giant. From
light travel time effects, it looks like this gas is out at a distance of one light
year. It will be amusing to see what happens ten years from now when the
debris hits the circumstellar fan.

The last point is nucleosynthesis. One strong clue is in the light curve,
which becomes very nearly exponential. You expect the explosion to cook
about 1/10 solar mass of ^{56}Ni. This doubly magic nucleus is created in the
inner part of the star, just outside the collapsed core. That nucleus decays
to ^{56}Co, which decays to iron by emitting gamma ray lines. Figure 14 shows
a straight line fit to the data. The observed rate of decline is very close to
the rate of decline that you'd expect from ^{56}Co. The flux that we see is the
flux you'd get if the star originally had seven hundredths of a solar mass of
iron, very close to what has been predicted. This may not sound like much,
but 20 000 seething Earth masses of radioactive debris can be the principal
energy source for the supernova after the initial shock heating.

The real test has come from seeing the gamma rays. That is as important
to the nucleosynthesis story as seeing the neutrinos was for core collapse.

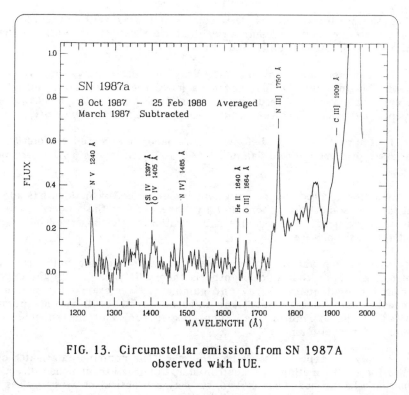

FIG. 13. Circumstellar emission from SN 1987A
observed with IUE.

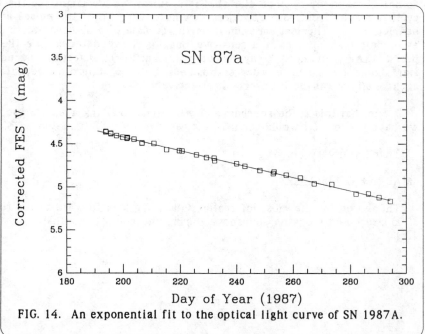

FIG. 14. An exponential fit to the optical light curve of SN 1987A.

Compton-scattered gamma rays coming out as x-rays, were seen by the Ginga satellite, and the gamma ray lines were reported both from the Solar Maximum Mission satellite and from balloon experiments.

To summarize, I think it's fair to say that the observations which we've been able to make on this supernova move the subject from indirect argument and plausible inference, the usual location of astrophysics, to experimental evidence. Each central idea has been tested in a new way.

Is it a massive star? Before 87A, we depended on stellar evolution theory; afterwards we identified Sanduleak -69 202, and it was a 20 solar mass object.

Was there core collapse with neutrino emission? This idea has gone from theory to something experimentally observed by the underground detectors. The energy, temperature and duration of the observed neutrino blast fit the basic picture.

Are supernovae the elemental philanthropists of the universe? We see strong evidence that nucleosynthesis theory is true from light curve, and from the high energy X-ray and gamma ray emission. Direct abundance measurements using optical and infrared spectra will become possible in 1988 as the expanding shell turns transparent and the supernova turns into a supernebula.

Finally, we expect to see a neutron star. The neutrino data match our picture for the formation of a neutron star, but this aspect needs direct confirmation. In the next year or two, we may see optical or X-ray pulses from a spinning, magnetic neutron star that is a pulsar, or we might see the energy that results from such an object slowing down even if the pulses are not beamed at us. Thermal emission from the surface of a neutron star is harder to detect and demands a sensitive imaging X-ray detector like the proposed AXAF (Advanced X-ray Astrophysics Facility). Detecting the neutron star should be the next order of business, but the supernova sets its own agenda: all we can do is observe it attentively.

I said that this is the supernova of a lifetime. Looking ahead to the work we have yet to do, I would say it is a supernova *for* a lifetime, too.

Thank you very much.

MAURICE GOLDHABER:

Thank you for the most interesting talk. We look forward to the results of measurements on the supernova during the coming months!
□□

MAURICE GOLDHABER:

We now come to the last talk of this meeting by Professor Paul Chu of the University of Houston. He was born in China, was educated first in Taiwan, then in New York City, and got a Ph.D. in San Diego, working with the famed, late Bernd T. Matthias, who had a special instinct for finding new supercondutors. Professor Chu seems to have inherited this and I'm very happy to introduce him now.

The Discovery and Physics of Superconductivity above 100 °K

Paul Ching-Wu Chu

Texas Center for Superconductivity

University of Houston

Houston, Texas 77204-5506

Thank you, very much. Coming to Cleveland is like coming home, since I have spent many years here.

Judging from my humble background, I am working as I should, within the noise level of the energy scale of the previous speakers, who were talking about high energy and astrophysics. In Houston, we're only working in the meV range, while they're working on 10 TeV, or even 10^{19} eV. In spite of that, I think physics in the low energy scale is quite interesting. Just like many of you, I feel very fortunate to be born at the right time, and working at the right place, and especially with the right group of colleagues. It is because of their dedication and hard work that I can be here talking about high-temprature superconductivity. They include P. H. Hor, R. L. Meng, Y. Q. Wang, L. Gao, J. Z. Huang, J. Bechtold, Y. Y. Xue, Y. Y. Sun, K. Burton, D. Campbell, T. Lambert, and A. Testa at the University of Houston, M. K. Wu and his group at the University of Alabama at Huntsville, C. Y. Huang at Lockheed, R. M. Hazen and his colleagues at the Geophysical Laboratory in Washington, D.C., J. Ho at Wichita Falls State University, and Y. C. Jean at the University of Missouri at Kansas City.

As is evident from Fig. 1, the field of superconductivity is not new; but it is still 25 years younger than the Michelson-Morley experiment. Superconductivity was discovered at 4.2K back in 1911 by Kamerlingh Onnes. Since then, people have realized many potential applications of superconductivity. However, there existed a temperature barrier due to the low T_c, the superconducting transition temperature. People worked very hard to raise the T_c without too much success until the 1930's, in Germany, where the study of ultra-hard material was very advanced. They raised the T_c up to above 10K in niobium carbide and niobium nitride, with a large critical current and a large critical field; in contrast to the earlier nontransition-metal superconductors, whose critical current and critical field had been very low. (Critical current and critical field are the maximum electrical current and the maximum magnetic field that a superconductor can tolerate without losing its superconducting properties.) After the 30's, people realized that the critical current barrier and the magnetic field barrier could be overcome, leading to the possible applications of superconductivity, only if one could overcome the temperature barrier. We all know that the practical high temperature is 77K, the boiling point of liquid nitrogen.

Though people were working very hard, there was little progress for

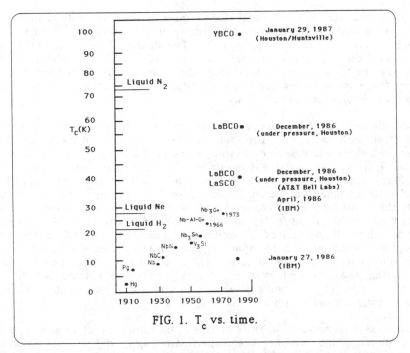

FIG. 1. T_c vs. time.

nearly 20 years. Then in 1953, Bernd Matthias, my former mentor, found superconductivity in Nb_3Sn at 18K. For the next 13 years, there was no advancement until 1966, when the temperature was raised to 21K. This was important because for the first time people could achieve superconductivity by the mere use of liquid hydrogen, which is plentiful, cheap, efficient and easier-to-handle, but not very safe. Seven years later, in 1973, John Gavaler and Lou Testardi observed superconductivity up to 23.2K in Nb_3Ge-films. For another 13 years nothing happened. Last year everything changed. When you look at this rate of increment, the T_c really moved up very slowly. If we assume linear progress, it would take more than 200 years to reach the practical temperature of 77K. I think none of us could live that long and even if we could, we would not have the energy to enjoy the excitement that many of us are enjoying today.

In general, we had three types of people. One group, the pessimists, which included some of the theorists, predicted that the highest T_c achievable was to be something like 30K. Another group, the optimists, think we should look for something else instead of following this traditional way of raising the T_c. Of course, the great majority is the third group: they just don't care.

The optimists, including Bernd Matthias, believed in the existence of high T_c and pushed very hard to get the T_c higher. Beginning in the 60's, people started looking for new ways to raise the superconducting transition temperature. In fact, in the early 70's, there were proposals for getting room-temperature superconductivity in several different systems, including organic systems. There was a conference held in Hawaii in the

early 70's called, "Room-Temperature Superconductivity in Organic Metals."
Bernd Matthias got so frustrated by the over-optimism expressed by some
that he wrote in Physics Today in 1971, "The deluge of idle speculation coming to us these days from all sides just won't do it; all it will manage to do is
to widen the credibility gap instead of the energy gap. In the spirit of our
times, there is an increasing tendency to substitute for nonexistent results
many words of great expectation."

Fortunately, I have a group of colleagues who share my dream of higher T_c's but with caution. Let me tell you one story of what happened in our
group, which is extremely dedicated, working very hard in raising T_c for
years. I remember in 1984 I dreamed of getting superconductivity at 77K
like Bernd Matthias used to do. Well, we all know that when we wake up, we
usually forget what we dreamed about. So, I tried very hard to remember
what the material was in the dream; it was a sodium-sulfur system. So I
went back to the lab next day, talked to my group, and told them about the
dream. They all got excited. They went to the library, found all the references, and made all the possible sodium-sulfur compounds in two weeks.
Then they tested them, and it turned out that, not only were they not superconducting, they weren't even conducting. Two weeks later, we dropped
the whole project, without any accusations or finger pointing. That's the
kind of spirit of the group.

Now, as I mentioned, in spite of this great effort trying to look in a new
direction for high-temperature superconductivity, there was no breakthrough until the publication, last year, of the paper, "The possible high-T_c
superconductivity in barium-lanthanum-copper-oxide," by Georg Bednorz
and Alex Müller, in Zeitschrift für Physik. The basis of their report was
their resistivity measurements on the lanthanum-barium-copper-oxide
system. They observed that the resistivity of this system started to decrease
at around 30K and dropped by several orders of magnitude at lower temperature. In some of the samples, resistivity actually went to zero at 12K. In
order to demonstrate this was indeed a superconducting transition, they increased the measuring current and found the transition to be suppressed to
a lower temperature, characteristic of a superconducting transition. But
the resistivity of the original samples was extremely high. Therefore
when one increases the measuring current, there can be a possibility to
generate local heating and to give rise to an apparent down shift of the
transition. Perhaps, this is why the report was received not very convinc-
ingly by people in the community worldwide, in as early as September
1986, as I found out later.

But, since we had been working on superconducting oxides for quite a
while, and we knew that oxides were funny animals, we felt that something
was right here. Within about a two-week period after we read the Swiss report of Bednorz and Müller in November 1986, we reproduced their results
although using a different technique to synthesize the compound, as
shown in Fig. 2.

In order to avoid possible local heating, we applied a magnetic field to
this material. You can see that the transition is shifted to a lower temperature, indicative of a superconducting transition. But, we know that some-

times we can see this kind of magneto-resistive effect in a normal-state material. So, for a further check, we measured the magnetic properties of the Swiss compound. It showed a diamagnetic shift below 20K (Fig. 3), demonstrating that superconductivity was indeed taking place in the compound, although it represented only a 2-3% Meissner effect in the compound. This was further confirmed by the I-V characteristic

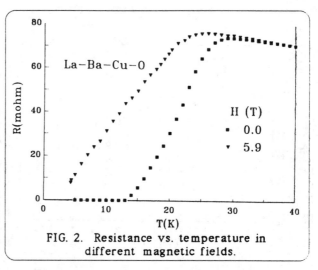

FIG. 2. Resistance vs. temperature in different magnetic fields.

measurements shown in the same figure. When the current exceeded the critical current density, the sample immediately returned to the normal-state. So there was no doubt in our minds that Müller and Bednorz indeed found superconductivity in the 30K range in this material.

When I first studied solid state physics, I read the book by Phil Anderson, *The Principle of Solids.* I learned two specifics: (1) a solid is defined as something when kicked, hurts your toes, and (2) all the interactions in a solid are electromagnetic in nature. We know electromagnetic interactions involve interatomic distances. So, we decided to use pressure as a means of exploring the nature of this high temperature superconductivity in this unusual class of materials. So we subjected the compounds to pressures.

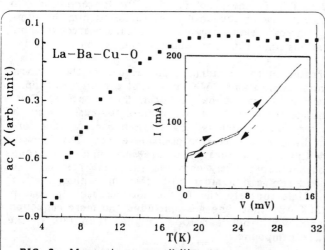

FIG. 3. Magnetic susceptibility vs. temperature. Current-voltage characteristics at 4K.

We found that T_c immediately rises very rapidly at a rate of a hundred times greater than an ordinary superconductor. This tells us something unusual indeed is happening. I should point out that all La-Ba-Cu-O samples examined up to this point were mixed phase. In December 1986, the Tokyo group had already identified the superconducting phase with a La_2CuO_4 structure, which is

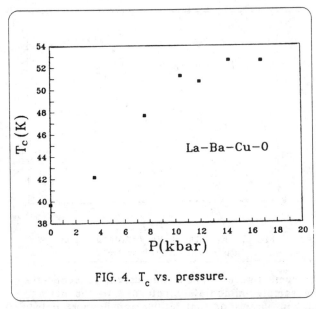

La−Ba−Cu−O

FIG. 4. T_c vs. pressure.

called the 214 phase, a ternary phase. Once the phase was known, using the chemical method by purifying the material, and at the same time using the physical method by squeezing it, we raised the T_c to 52.5K, in mid-December 1986. Within another two weeks, by the end of 1986, we pushed the temperature up to almost 57K, in the La-Ba-Cu-O 214 compound, as shown in Fig. 4.

This was significant because it gave us confidence that superconductivity at much higher temperatures must be possible. Before that time, the existing theory told us that superconductivity could not exceed 30K with a 10K uncertainty. The theorists were right for all that time because experimentalists for years couldn't get the temperature higher than 30K. But this new data proved the contrary. In fact, as soon as we found that pressure could raise the T_c of La-Ba-Cu-O at a rate of 1K/kbar, I immediately called my friend Dave Mao at the Geophysical Laboratory. He had developed a high-pressure diamond cell which could be held in the palm of his hand. By turning only a little screw, he could generate megabars in his palm. Both he and I felt that perhaps T_c could be raised to 1,000K, because one million bars would give us about a 1,000K increase in T_c. So the problem would no longer be how to create superconductivity, but how to quench it. Before I sent him the sample, we found that the T_c-increase slows down when pressure exceeds about 12 kbar. To save his precious time, I did not send him the sample.

The public got interested in this material, not because of the science, but because of its potential applications. We knew that the physical application of pressure to raise the T_c is not that practical. So we decided to use chemical methods to simulate the pressure effect. One of the easiest ways is to replace barium with strontium which is smaller than barium, therefore reducing the interatomic distance just like the pressure effect. The T_c went from 30K up to 42.5K without the application of pressure. The Bell group and the Tokyo group both independently obtained the same results. We took the step further and replaced barium with calcium, which is even smaller than strontium. Unfortunately, the T_c did not increase. Instead, it dropped back down to 20K. By that time we concluded the mere reduction in the interatomic distance would not lead to T_c higher than 60K in the Swiss compound or the 214 compound.

By examining the large volume of data accumulated by the end of 1986, we noticed the existence of sharp drops of resistivity at temperatures above 70K, indicative of a superconducting transition. But this only happens in compounds with multiphases, including phases in addition to the 214 phase, telling us that, first, a T$_c$ higher than 70K may be possible, and second, if we are going to look for high T$_c$, we have to get away from the Swiss compound.

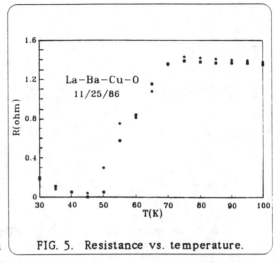

FIG. 5.   Resistance vs. temperature.

In Fig. 5, I show you what I meant by seeing an indication of superconductivity. Back in November 1986, for example, we saw the resistivity start to drop very rapidly around 70K. Unfortunately, the drop disappeared in a few days. Since we did not know the exact composition nor the structure of a really high-temperature (i.e. > 70K) superconductor, we decided to purposely make samples with a distribution of compounds in each of them. A barium rich La-Ba-Cu-O sample was then synthesized in a reduced atmosphere. The sample was pink and insulating outside. It was no longer suitable for resistive check. So, we decided to do magnetic measurements. The results are shown in Fig. 6 (curve a). The magnetization started to show a diamagnetic shift at about 100K. At 4.2K we saw a Meissner effect of approximately 40% of a bulk superconductor. This happened on January 12, 1986. With such a large diamagnetic signal, there was no doubt in our minds that superconductivity above 90K must exist. Finally, we all had a day of rest. The students went home and had a nice sleep. The next day they came back and tried to clear up the problem. But they cleared up the signal instead of the problem and the whole diamagnetic signal disappeared, as you can see in the same figure (curve b). In spite of this disappointing observation, we learned that superconductivity above 90K must exist. The only question then was how

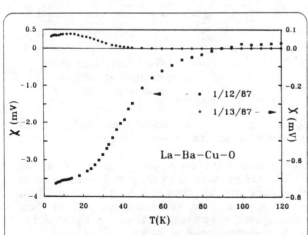

FIG. 6.   Magnetic susceptibility vs. temperature on two different days.
(Superconductivity above 77K unambiguously established for the first time.)

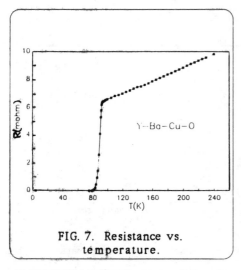

FIG. 7. Resistance vs.
temperature.

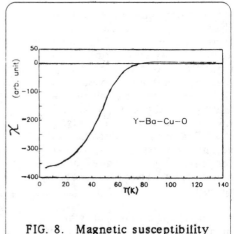

FIG. 8. Magnetic susceptibility
vs. temperature.

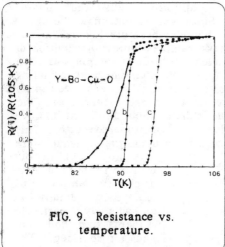

FIG. 9. Resistance vs.
temperature.

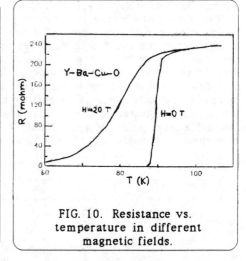

FIG. 10. Resistance vs.
temperature in different
magnetic fields.

to stabilize it. From the high-pressure data, we know that smaller atoms may help, so we immediately replaced lanthanum with yttrium.

Finally, on January 29, we obtained the curve shown in Fig. 7, where resistivity of the Y-Ba-Cu-O (YBCO) sample started to drop at 93K and goes to zero at around 80K. In order to show that the drop is indeed associated with a superconducting transition, we measured the magnetization and, as indicated in Fig. 8, it showed a diamagnetic shift below 90K with a signal corresponding to about 24% of a bulk superconductor at 4.2K.

There was no doubt that superconductivity above 90K had been stabilized and observed unambiguously. Within a few days we sharpened the transition and then pushed the onset T_c up to 98K (Fig. 9). This T_c remains the current record of stable and reproducible superconductivity.

To determine the magnetic barrier, we measure the resistivity of YBCO in the presence of magnetic field. It is clear from Fig. 10 that, even at 20 Tesla, it is still superconducting above 60K. Based on this data, using extrapolation, one could get an upper critical field up around 200T. In other words, it requires two million gauss of magnetic field to quench the superconductivity to 0°K, in contrast to the 0.5 gauss of the earth's magnetic field. Based on this kind of upper critical field, one could get the coherence length, approximately 15Å, which is unusually short for superconductivity.

The next problem was to determine the structure of this material in collaboration with the Geophysical Laboratory at Washington, D.C. Both the stoichiometry and the basic atomic-arrangements (except for some oxygen atoms) of the YBCO compound were obtained in late February 1987. In Fig. 11, I have shown the structure of YBCO refined by the Argonne group. Basically, it has three copper-oxygen layers per unit cell. The central layer is formed of copper-oxygen chains. Its composition can be represented by the formula $YBa_2Cu_3O_7$-d and because of the composition it's called the 123 compound.

I should point out that this is the first true quaternary compound superconductor ever discovered, and the 30K superconductivity exists in the basic ternary compound La_2CuO_4, by partially replacing La with Ba, Sr, or Ca. In other words, this new 123 structure requires four elements to stabilize it, unlike the 214 ternary structure, which needs only three elements to form.

Once the structure and the stoichiometry are known, the next problem is to determine what the active elements responsible for this high-temperature superconductivity are. People working in the field all know to use the magnetic elements to probe superconductivity. The easiest thing to do is to replace yttrium with the magnetic rare earth elements to see how the two competing phenomena -- superconductivity and magnetism -- affect each other. So we did that. Instead of seeing a ferromagnet with the complete replacement of yt-

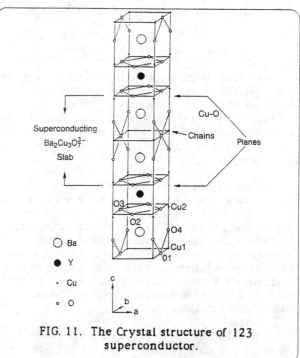

FIG. 11. The Crystal structure of 123 superconductor.

trium by gadolinium or europium as expected, for example, we observed superconductivity with a T_c totally unsuppressed. Within a few weeks we found that the replacement of yttrium with almost all of the rare earths (except Ce, Pr, and Tb) does not perturb the superconductivity. A new class of superconductors $ABa_2Cu_3O_7$-d, with A = La, Y, Nd, Sm, Eu, Gd, Dy, Ho, Er, Tm, Yb, and Lu, was thus discovered. It should be mentioned that groups at Brookhaven/Ames, Bell Labs, Los Alamos, and Tokyo had also independently observed similar results.

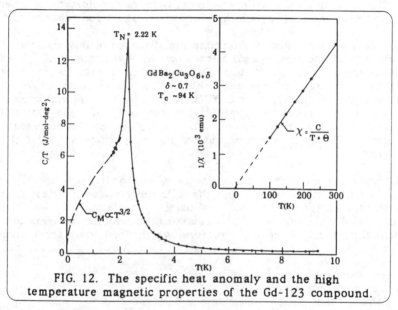

FIG. 12. The specific heat anomaly and the high temperature magnetic properties of the Gd-123 compound.

Our observation immediately led us to suggest that superconductivity really occurs in the three-layer system and that the yttrium and the rare-earth elements are there only to stabilize the structure. The rare-earth elements are not participating actively in the superconducting process. By comparing the 214 and 123 compounds, we also concluded that higher T_c should be achieveable by looking for compounds with a greater number of Cu-O layers.

In addition to this high-temperature superconductivity, I would also like to show that we have also observed magnetic ordering at low temperature due to the long range order of the rare-earth magnetic ions. For example, Fig. 12 displays the results of $GdBa_2Cu_3O_7$-d. Jim Ho at Wichita State University, in collaboration with us, found that the specific heat shows a λ anomaly at about 2.2K, indicating that an antiferromagnetic ordering is coexisting with superconducting ordering in consistency with our earlier magnetic data shown in the same figure. So the system provides us with a very nice basis for the study of the interaction between superconductivity and antiferromagnetism.

I would also like to point out that the yttrium reserve is not that great on earth. If only yttrium works, a full-scale application of high tempera-

ture superconductivity is not really practical. Lanthanum is plentiful, especially in this country. Fortunately, we immediately got lanthanum 123
superconducting above 90K in March 1987. However, we still have a problem with lanthanum because the lanthanum 123 compound gives us a T_c
starting at 92 to 95K, but then the resistivity does not go to zero until 75K.
This is related to its large atomic volume which we have to overcome. So it
still needs some more work, although we have already developed a gas-ex-
change technique which occasionally gives a temperature above 90K for
the resitivity to vanish.

Extensive efforts have been devoted to the study of high temperature
superconductivity over the last nine months and great progress has been
made. Now, let me first summarize below what we now know about this
class of materials:

1.	transition temperature	90 - 100K
2.	upper critical field	150 - 200 T
3.	coherence length	1.6 - 30Å
4.	critical current denisty	$10^3 - 10^6$ A/cm^2
5.	carrier concentration	10^{21}/cm^3
6.	density of states	(3-7) mJ/mole K^2
7.	universal gap equation	$2/kT_c \sim 1.3 - 20$
8.	glassy behavior	
9.	layer-like structure	
10.	superconductivity mainly confined in three Cu-O layers	
11.	coexistence of superconductivity and antiferromagnetism (due to Cu and/or rare-earth ions)	
12.	electron pairing	
13.	substantial fraction of electrons participating in superconductivity	
14.	no isotope effects in 123; partial in 214	
15.	superconductor -- semiconductor -- magnet	

We know that the transition temperature is between 90K to 100K. I'll
comment on the recent reports on higher T_c later. The upper critical field
is between 150T and 200T and the coherence length is 1.6 to 30Å, depending
on who's data you're using. The critical current density for the bulk material remains at 10^3 A/cm^2. But thin films, properly made, have a current
density up to 10^6 A/cm^2. The carrier concentration is very low for this
class of material and the net carriers are holes; it's almost one or two orders
of magnitude lower than conventional superconductors with a high T_c, like
20K. Then the density of states for this class of material is extremely low.
The low density of states together with the high T_c puts the 123 and 214
compounds in a different category. The universal gap equation, at this moment, is still debatable; the $2D/kT_c$ is 1.3 to 20, although they tend now to
converge to 7, in contrast to the BCS prediction of 3.5. People have also observed glassy behavior in this material, meaning the magnetic flux moves
slowly after you change the field or temperature, i.e., it takes time for the
flux to move out and move in. The 123 compounds have a layered structure
shown earlier. The superconductivity, as concluded from the rare-earth
substitution experiments mentioned earlier, occurs in the three Cu-O layer
assembly. Then we also know that in this class of material when you get

the oxygen out, expecially in the 30K material, one can also see antiferromagnetic ordering due to the Cu-ions. When the antiferromagnetic ordering disappears with increase of oxygen content, superconductivity appears. Tunneling and microwave experiments show that electrons in this
class of material pair when they are in the superconducting state. In addition, a substantial fraction of the electrons are participating in superconductivity as indicated by the positron annihiliation experiments; I'll come
back to this point later. A negligibly small isotope effect was observed in
the 90K 123 superconductor, but a partial isotope effect was observed in the
30K 214 superconductor. This material can be made into a superconductor,
or a semiconductor or a magnet without changing the chemical elements,
except the oxygen content.

Now what don't we know? We have more things that we don't know
than we know, of course. Let me first list them:

1. superconducting mechanisms -- phonons, bipolarons, plasmons, para
   magnons, negative U-centers, charge fluctuations, spin fluctuations,
   resonating valence bonds, or spin bag?
2. dimensionality -- 1, 2, or 3?
3. valence states of copper -- +1, +2, +3 or mixture?
4. implication of partial isotope effect -- role of phonons?
5. glassy behavior -- intrinsic or extrinsic?
6. superconductivity vs. magnetism -- accidental or consequential?
7. superconducting transition vs. metal-insulator transition -- accidental
   or consequential?
8. roles of oxygen and defects -- stoichiometry, symmetry, or linear
   chain?
9. stability -- thermal cycling, environment, electric field, stress and
   strain, etc.?

This morning, Phil Anderson discussed one of the possibilities. Others
have also been proposed: like phonons, bipolarons, plasmons, and so forth.
At present, there exists not enough experimental data to favor one model
over the other unambiguously, although recent data does point out the importance of the magnetic interaction, due to Cu-ions, as advocated by
Anderson. The dimensionality of superconductivity in this class of material is not quite clear. Some people proposed that superconductivity occurs
in the linear chain in the center layer of copper-oxygen and others think
that the copper layer is responsible for superconductivity. However, we
believe that the assembly of three layers of copper-oxygen is the answer
and the 123 compound should be a highly anisotropic three-dimensional
superconductor. The question concerning the valence states of the copper
remains unclear at this moment. Many of the models depend heavily on
the valence state of Cu-ions, even though the concept of valence becomes
vague when one deals with a metal, such as the high T_c superconductors,
where the concept of conduction bands prevails. For example, based on
similar data from near-edge X-ray absorption experiments, people at
Brookhaven said it was only +2 and those at Argonne said it was +2 and +3,
and so on. A similar situation arises in the interpretation of the XPS data.
As for the resonance data, some saw the ESR line and attributed it to the
Cu^{+2}-ions but others didn't see any, depending on the samples. Of course,

the absence of the ESR Cu^{+2} -line can be due to reasons other than the absence of Cu^{+2} -ions, e.g. local short-range magnetic order.

The next question is the implication of the partial isotope effect in the role of phonons in this class of material. People usually take the absence of an isotope effect as evidence for the absence of a phonon role in superconductivity. However, we do know there exists no isotope effect in some of the conventional low-temperature superconductors such as Ru, and yet the superconductivity can still be accounted for by electron-phonon interaction. Next comes the question of the glassy behavior of the 123 superconductors. If it is intrinsic, is it related to the specific structure of the material or just the twin boundary due to the tetragonal-orthorhombic transition? If it is intrinsic to the 213 and 123 superconductors, we shall have a problem for applications of these materials. Magnetism and superconductivity are known to occur in this class of material although at different oxygen concentration ranges. Do they occur accidentally or is one the consequence of the other? Superconductivity occurs in this class of material, which can also be made into an insulator. Therefore the question also arises whether the occurrence of the superconducting and metal-to-insulator transitions is accidental or is one the consequence of the other?

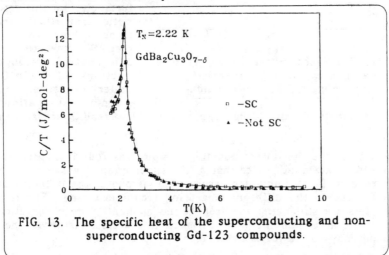

FIG. 13. The specific heat of the superconducting and nonsuperconducting Gd-123 compounds.

The role of oxygen has been determined to be very effective in changing the superconducting properties in this class of material. Whether it affects superconductivity through stoichiometry, through symmetry, or through linear chains, or others, remains unknown. The stability of this material is unknown. It varies with thermal cycling, environmental conditions, electric fields, stress and strains, and so on. On the other hand, we do know if the compound is made more ideal, the more stable chemically it becomes.

Now, in the face of all these unknowns, people have carried out many experiments in search of answers. For example, in order to probe the superconductivity one can put magnetic impurities into it and see the response of its superconductivity. Now we can do the opposite and see how

magnetism responds to superconductivity. What we have found is shown in
Fig. 13. When it is superconducting we can see the antiferromagnetic ordering in the Gd-Ba-Cu-O 123 compound evidenced by the specific heat
anomaly, at 2.2K. Then we pumped the oxygen out and made the sample
nonsuperconducting. An identical specific heat anomaly to that in the superconducting sample was detected in the nonsuperconducting sample.
This tells us that superconductivity and the antiferromagnetism due to the
magnetic rare-earth element (not Cu-ions) are not coupled to each other.

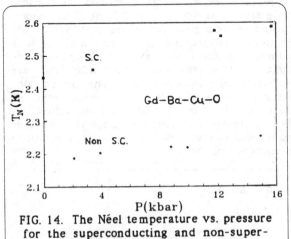

FIG. 14. The Néel temperature vs. pressure
for the superconducting and non-superconducting Gd-123 compound.

But then, what would hap-
pen if we start changing
the lattice parameter as one
would do in an attempt to
raise T_c, for example?
Would this non-participa-
tion of the rare earth elements in the superconducting process still be valid?

Therefore, we decided to
determine the pressure effects on the antiferromagnetic transition tempera-
tures, in both types of samples. We measured the specific heat of the superconducting Gd-Ba-Cu-O 123
under pressure, and the
magnetic susceptibility of
the non-superconducting Gd-Ba-Cu-O sample. The results are summarized
in Fig. 14.

The T_N of the nonsuperconducting sample was found to increase with
pressure at a rate about twice that of the superconducting sample. There's
a difference in T_c at ambient pressure. This is due to the different definition of T_N, due to the different measuring techniques used. The implications of this are still being analyzed at the moment. The observation can
imply atomic arrangements with different interatomic distances, the rare-
earth atom may not be just a by-stander.

In an attempt to determine the location of superconductivity micro-
scopically, we look into positron annihilation in this material in collaboration with Jerry Jean at the University of Missouri at Kansas City. You can
see from Fig. 15 that the positron lifetime in the YBCO 123 poly-crystalline
sample drops very rapidly at T_c, in contrast to conventional intermetallic
superconductors. This is very unusual. The Doppler broadening parameter
also exhibited an anomaly at T_c. In collaboration with C. K. Sun at Langley,
we have also observed a drop in the acoustic attenuation at T_c which is in
the wrong direction, as predicted from thermodynamic considerations.

Anomalies in the acoustic velocity have also been reported by Bishop at
Bell Labs and Migliori at Los Alamos.

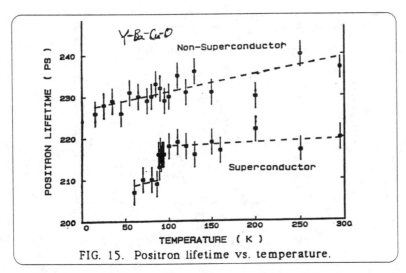

FIG. 15. Positron lifetime vs. temperature.

Upon cooling, the velocity goes up, in the wrong direction from prediction. In addition, the acoustic velocity jump is almost two or three orders of magnitude larger than the thermodynamic prediction.

A high resolution X-ray study by Paul Horn at IBM displays a drastic change in the orthorhombicity of the material at T_c.

So then, the question is the following: what is the implication of all these anomalies at T_c, as listed below:

1.   a superconducting transition;
2.   a positron lifetime anomaly;
3.   a Doppler broadening parameter anomaly;
4.   an orthorhombicity anomaly;
5.   a large sound velocity increase;
6.   a large acoustic attenuation drop.

Since the thermodynamic predictions for a superconducting transition have to be correct, the anomalies at T_c in the acoustic measurements are puzzling. The problem can be resolved if one assumes that there exists another transition coinciding with the suprconducting transtion. This is consistent with the Doppler broadening parameter (indicative of a change in the electron energy spectrum), and the orthorhombicity anomaly. If the suggestion is true, the study of the relationship between the new phase transition and the superconducting transition may shed light on our understanding of high temperature superconductivity.

We all know superconductivity depends sensitively on the oxygen content. Let us take the europium barium copper-oxide 1-2-3 compound as an example. Originally $EuBa_2Cu_3O_7$-d had a T_c around 90K. When we pumped on it at low temperature the Tc, all of a sudden, switches from 90 to 60K before it decreases to zero, as shown in Fig. 16. In other words, T_c exhibits a flat step as a function of oxygen concentration near d - 0.4 as shown in the

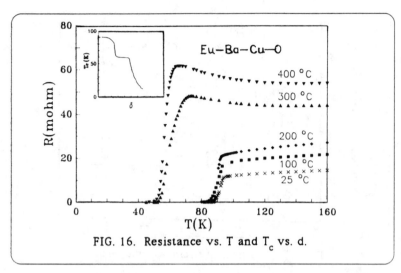

FIG. 16. Resistance vs. T and T_c vs. d.

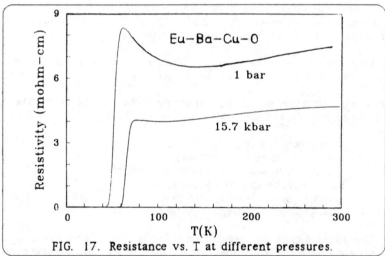

FIG. 17. Resistance vs. T at different pressures.

insert of Fig. 16. Some people suggested that the 60K-material represents a
new phase, but others not. We decided to examine the 60K-compound in an
attempt to understand the occurrence of high T_c superconductivity, by
searching for similarities and dissimilarities between the 60K- and 90K-
$EuBa_2Cu_3O_7$-d samples. Fig. 17 displayed the resistivity of the 60K-Eu-Ba-Cu-
O 123 compounds at ambient and high pressures. At ambient pressure, the
resistivity first drops almost linearly on cooling, then increases, indicating
some kind of localization before superconductivity sets in near 60K. When
we apply pressure to the 60K-sample, we can see that the pressure shifts
the T_c rapidly upwards and then destroys the resistivity anomaly immediately above T_c. At the same time, the room-temperature resistivity is rapidly and drastically suppressed as shown in Fig. 18.

At about 12 kbar, the resistivity drops by almost 40% in this material. This is in strong contrast to the 90K-sample, for which pressure enhances the T_c only slightly and suppresses the resistivity only slightly. The rate of T_c-enhancement by pressure is about 7 to 10 times greater than the 90K-sample, almost the same as the 30K-material.

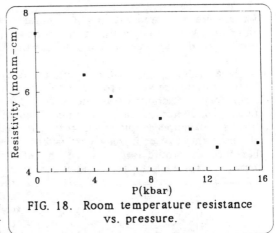

FIG. 18. Room temperature resistance vs. pressure.

The simultaneous occurrence of superconductivity and antiferromagnetic ordering in the 214 and perhaps also the 123 compounds depending on the O-stoichiometry, prompts many to suggest the important role for the antiferromagnetic correlation in high-temperature superconductivity. If one assumes that antiferromagnetism in the system stems from the super-exchange interaction with a dominant kinetic part, it can be obtained that the Néel temperature T_N - zt^2/U and dT_N/dP - $(10/3)(T_N/B)$, where z is the number of the nearest magnetic neighbor ions, t the transfer integral, and U the intra-atomic Coulomb repulsion. The relations have been demonstrated by Kaneko to describe well the experimental data including the antiferromagnet of slightly oxygen-deficient La_2CuO_4. Based on a Hubbard Hamiltonian including the antiferromagnetic interaction, Cyrot showed that T_c - t^2/U. The similar dependence of T_N and T_c on t^2/U led to the suggestion that dT_c/dP may also follow the same dependence and be equal to $(10/3)(T_c/B)$ as T_N, or $\phi \equiv (B/T_c)(dT_c/dP)$ - $(10/3)$. The following values are obtained by us for ϕ: 3.2(90K-YBCO, 90K-EBCO); 30(60K-EBCO); and 14-40(30K-LSCO, 30K-LBCO). The results suggest that the 60K-superconductivity may be more similar to the 30K-one than the 90K-superconductivity.

Let me summarize some of the properties of the 30K-214 and 90K-123 compounds below:

	30K-superconductivity	90K-superconductivity
Crystal Chemistry	Ternary	Quaternary
Structure	Single-layer	Triple-layer
dT_c/dP (K/kbar)	1 - 0.8	0.1 - 0.2
$\alpha(T_c \propto M^{-\alpha})$	- 0.2	- 0.01
e^+ Lifetime at T_c	slight increase	drastic drop
Acoustic Softening	yes	no

It therefore appears that the 30K-material is quite different from the 90K-material. Now the remaining question is about the 60K-material: does it form a distinct class or does it fall into the 30K-material class?

Finally, let me make a few comments on the possibility of very high T_c. First, I would like to borrow some of the definitions from Argonne National Laboratory. A conductor should look like Fig. 19. For a semiconductor, you cut the conductor in half. A superconductor is a conductor in a Superman suit. A 90K-superconductor is obtained when you bend him over 90°. We all know this is not a very comfortable position to be in for a long time, therefore we would like to have an 180K-superconductor standing upside-down. In spite of the fact that we're working on a very low energy scale, the pressure is pretty high. Such a position could still give us trouble. So, the most comfortable position is 360K-superconductor. That's the reason we're working hard on raising the T_c.

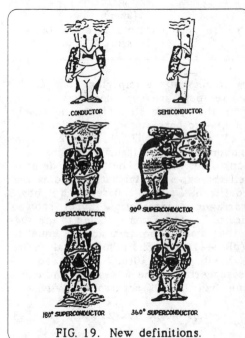

FIG. 19. New definitions.

Of course, we also know that scientists already have enough problems with the 90K-superconductivity. High T_c will pose even more challenge to scientists. From the application point of view, the optimal operating temperature for a large-current superconducting device is about 70% of the T_c. In other words, if we want to operate a large-current superconducting device, we need a superconductor with a $T_c > 110K$. As Phil Anderson mentioned this morning. there were many indications of superconductivity reported by different groups all over the world. They started out from Houston, then Berkeley, Japan, Wayne State, Stanford, Energy Conversion Devices, Lockheed/Houston, China, India, Yugoslavia, Colorado State, North Carolina State, Maryland, and Phillips North America. I just heard a rumor from Georgia Tech at about 500K a few days ago.

Typical indications of superconductivity at temperatures substantially higher than 90K are shown in the next few figures. At Wayne State, J. T. Chen observed a dc voltage across a multiphase YBCO sample when it was exposed to a microwave source and cooled below 240K (Fig. 20). He attributed the observation to the so-called reverse Josephson effect due to an array of Josephson junctions formed by a mixture of the nonsuperconducting and superconducting components in his sample.

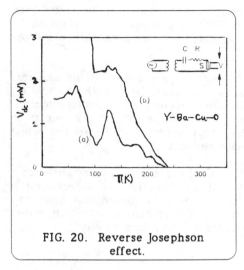

FIG. 20.   Reverse Josephson
effect.

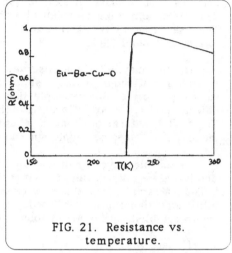

FIG. 21.   Resistance vs.
temperature.

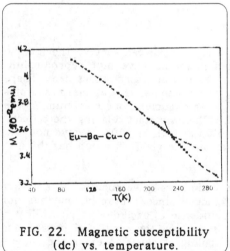

FIG. 22.   Magnetic susceptibility
(dc) vs. temperature.

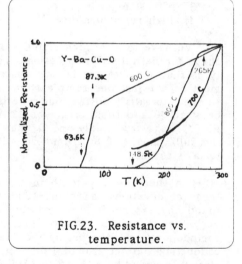

FIG.23.   Resistance vs.
temperature.

Large resistivity drops were observed by several different groups, for example by C. Y. Huang and us in Eu-Ba-Cu-O (Fig. 21). Resistivity goes down
to almost to zero at 225K.

The Magnetization measurement of the same sample shows a sharp
change in slope (Fig. 22), indicating there may be a small component of
diamagnetism. The combination of these resistive and magnetic results is
suggestive enough that superconductivity at high temperatures may exist,
but is certainly not yet proved. This evidence is definitely not sufficient. I
should mention that no efforts of ours so far could reproduce the observations in another sample.

In China, at Beijing Normal University, they implanted their YBCO 123-

samples with flourine ions. The superconductivity was first destroyed (Fig. 23). After annealing between 600 and 800°C, superconductivity came back. They saw resistivity starting to drop at very high temperatures and going to zero at around 150K.

In summary, all these observed anomalies are not very stable and this is why I call them unstable superconducting anomalies, in short, "USA's," (to be patriotic). Some people like to call them unstable superconducting objects, or "USO's," to differentiate them from unidentified flying objects or UFO's. But, it turns out that in Japanese, "USO" is pronounced *"oo-so,"* which means *"a lie"* to some people who do not believe it.

In the old days, when we tried to establish the existence of superconductivity, we only required zero resistivity and the Meissner effect for the bulk of a sample. For this new class of materials, we cannot take high stability and high reproducibility for granted. Therefore, we need four criteria to establish the unambiguous existence of superconductivity:

1. zero resistivity
2. Meissner effect
3. high stability
4. high reproducibility

We have to have high stability so the material can survive thermal cyclings for definitive examination. Then it should have high reproducibility so that, not just one laboratory can see it, but many others can. There must be not *just one* sample showing these anomalies, but many samples. In order to satisfy all the four criteria, at this moment, our maximum T_c lies between 90 and 100K, or maybe 120K. When we start relaxing the criteria, the T_c rises very rapidly and T_c may go up as high as 500K. I have not read the 500K paper yet, but rumor says that the Georgia Tech group has the material stabilized.

Can there be any truth in these anomalies? Personally, I feel that these observations cannot be totally ignored since presently, one has not found any experimental or theoretical evidence to exclude the possibility of very high T_c unequivocally. Let me show you the results of one of our compounds (Fig. 24). During the first cooling, it starts to enter the superconducting state at around 130K. In the subsequent warming, the onset T_c already dropped back down to 110K (Fig. 24, Run B). The next day (Fig. 24, Run C) the transition stayed at the same temperature. There was not much change afterwards. Something unstable was clearly happening here. This shows that cooling is bad for superconductivity. So far, we have talked about only the onset T_c. I would like to show you the cooling effect on the temperature at which resistivity goes to zero. It is equally bad for the case shown in Fig. 25.

The first time, when we cooled it down, the resistivity went to zero at 104K; but the temperature was already down to 102K on subsequent warming. On the second cooling, it was 98K; and finally, it moved toward and stabilized at 93K. It turned out that cooling can also be good for high-temperature superconductivity. We had a sample in which the resistivity showed

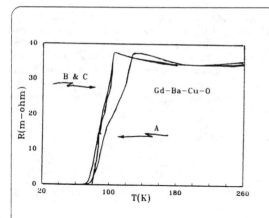

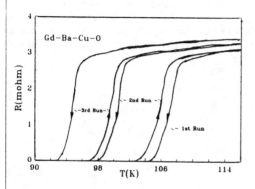

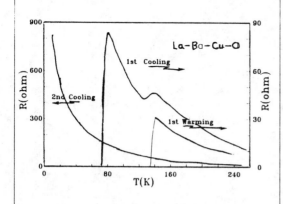

FIGS. 24, 25 and 26.
Resistance vs. temperature during
different thermal cyclings.

a drop at 140K and went to zero at 77K upon the first cooling (Fig. 26). When it was warmed back up to 120K, we had still not detected any resistance. A careful check at this temperature showed that there was no problem in the electrical circuit. Continued heating brought the sample back to the normal state at -130K, but unfortunately, when we cooled it down again, it became semiconducting and superconductivity was gone.

So, with all that, I believe superconductivity at higher temperatures is possible. At this moment the most important problem to overcome is the poor stability and the poor reproducibility of the anomalies. If the anomalies are indeed proven to be associated with superconductivity, the superconductivity must be filamentary in nature due to phases similar or related to the 123 compounds, independent of the lanthanide elements. However, we found that no such anomaly has yet been detected in the stoichiometric and ideally-made 123 compound.

The history of recent developments shows us the 30K material was not stable at the beginning, and neither was the 90K material until the material was isolated. We don't know at this moment if history will repeat itself. But I believe that the possibility exists, because I strongly feel that nature has been very nice to us for us to be able to discover 90K-superconductivity, but I don't believe it is so kind that we will be able to hit the biggest jackpot the first time.

In conclusion, in the science area, there will still be a lot of scientific problems and challenges lying ahead. Higher T_c is a real possibility, and known applications pose great promises and challenges as well to all of us. But I think the area of novel applications tailored to the unusual properties of this class of materials will hold even greater promise. Indeed, the year of 1987 is a super year in physics. We have witnessed superconductivity, supernova, superstring, and superconducting supercollider. Let us enjoy all these super-events in physics.

Thank you very much for your attention.

MAURICE GOLDHABER:

Thank you very much, Professor Chu. It's necessary also to have great enthusiasm, and you have it.

Well, this ends the formal proceedings. It's up to me now, for all of you, to thank the two organizers of this scientific session, who were able to attract such busy superstars. Bill Fickinger and Ken Kowalski, of Case Western Reserve University, have done a superb job in getting this meeting organized and run and we want to thank them very much. Thank you all for coming.□

Registered Participants
Modern Physics in America Symposium

Issam R. Abdel Raziq	Ohio University
Amir M. Abduljalil	Ohio State University
Sugath Abegunrathna	Kent State University
Joan Acker HM	Borromeo College of Ohio
Robert Acquarelli	University of Michigan
E. Merrill Adams	Case Western Reserve University
Kassa Adel	University of Michigan
Moutaz Adham	Bethany College
Nilanjan Adhikari	University of Pittsburgh
Stephen L. Adler	Institute for Advanced Study
Mike Agusta	Bethany College
C. J. Ahola	Cleveland, OH
Robert Akins	Case Western Reserve University
Naman N. Al-Niemi	Ohio University
John D. Alexander	University of Akron
Bill Alexander	Bethany College
Abdo Al Fakih	University of Michigan
Gabriel Allen	University Heights, OH
David Allender	Kent State University
Martin Alperz	Moreland Hills, OH
Ali Alramadhen	University of Michigan
Miltiadis S. Anagnostou	Ohio State University
Richard Andaloro	University of Toledo
Frederic J. Anders, Jr.	Wilmington, DE
Philip W. Anderson	Princeton University
Philip Anderson	Cleveland Heights, OH
Merrill L. Andrews	Wright State University
David M. Anthony	York University
Eugene F. Apple	General Electric Co.
Like Arabpour	College Park, MD
Robert L. Armstrong, Jr.	Akron, OH
Richard A. Arndt	Blacksburg, VA
Fred Arnold	Bowling Green State University
Henk Arnoldus	State University of New York, Buffalo
Heide Arnoldus	State University of New York, Buffalo
Paul R. Aron	NASA-Lewis Research Center
Neil W. Ashcroft	Cornell University
Gordon J. Aubrecht	Ohio State University
David H. Bachman	University of Akron
Roger Bacon	Bay Village, OH
Charles R. Bacon	Ferris State College
Waris M. Baig	Youngstown State University
Kenneth T. Bainbridge	Harvard University
Dan Baize	Western Kentucky University
Gust Bambakidis	Wright State University
Juilo Gea-Banacloche	Miami University
Amitava Banerjea	S. Euclid, OH
Shampa Banerjee	Case Western Reserve University
Xue-Jun Bao	Carnegie-Mellon University
Rachael Barbour	BP-America Research
David H. Barbour	Shaker Heights, OH
Michael P. Barrick	Parma, OH
Fernando Barrios	University of Cincinnati
Bill Barry	Hellerton, PA
John Bates	Mentor, OH

Barbara Bates	Lakeland Community College
James Batman	Wooster, OH
Andrew Bazarko	New York, NY
Mrs. Elmer G. Beamer	Cleveland, OH
Elmer G. Beamer	Cleveland, OH
George B. Beard	Wayne State University
Roger Becket	University of Dayton
James Beecher	Gencorp, Inc.
R. Begun	Cleveland Heights, OH
S. J. Begun	Cleveland Heights, OH
Jeffrey S. Beis	University of Toronto
Joseph Belak	Avon, OH
Elwyn Bellis	Marshall University
Jean Bennett	University of Alabama
Theresa Benyo	Kent State University
David Bergman	Ohio State University
Paul Bergstrom	University of Pittsburgh
Melissa Bernard	Avon High School
Wes Berseth	York University
Ralpj C. Bertonaschi	Cleveland, OH
Hans A. Bethe	Cornell University
Kumar Bhatt	Western Kentucky University
Alok Bhattacharyya	University of Cincinnati
Sourav Kumar Bhunia	Case Western Reserve University
Gary F. Bianchini	Case Western Reserve University
William Bidelman	Case Western Reserve University
Joseph R. Biel	Fermi National Accelerator Laboratory
Nelson M. Blachman	GTE
Christopher Blasko	Case Western Reserve University
Frank J. Blatt	University of Vermont
R. P. Blonski	Case Western Reserve University
Julie Blum	University of Michigan
Fernando Boada	Case Western Reserve University
Eric Bobinsky	NASA-Lewis Research Center
Eric Boerner	Case Western Reserve University
John Bognar	Wright State University
David Bogrees	Wright State University
Jerry Bohinc	Gates Mills, OH
Henry V. Bohm	Wayne State University
Randy Bohmer	Bethany College
Scott Bonham	Case Western Reserve University
Daniel Boyne	Case Western Reserve University
Robert I. Boughton	Bowling Green State University
Xavier Boutillon	University of Michigan
D. R. Bowman	University of Akron
Darwin Boyd	Kent State University
Frederick P. Boyle	Imperial Clevite Tech Center
Donald W. Boys	University of Michigan
Michael Bradie	Bowling Green State University
Wendy A. Brandts	University of Toronto
Clyde B. Bratton	Cleveland State University
David Braun	York University
J. F. Breen	General Electric Co.
Michael Bretz	University of Michigan
Margaret Bretz	Ann Arbor, MI
David Brick	Case Western Reserve University
David Brinker	NASA-Lewis Research Center
William Brito	University of Michigan
Robert L. Brott	Pennsylvania State University
Daniel C. Brown	Cleveland, Ohio
Glenn Brown	Case Western Reserve University

Laurie Brown	Northwestern University
Robert Brown	Case Western Reserve University
Stanley A. Bruce	Notre Dame University
Margaret Bruening	CEI. Perry Power Plant
Polly Bruner	Cleveland. OH
Clark Bruner	Cleveland. OH
Steven W. Bucey	Kent State University
Frank Bynum	Miami University
Zhi-xiong Cai	Michigan State University
Jin Cai	Bowling Green State University
Brent Campbell	Case Western Reserve University
Gregory Canfield	Case Western Reserve University
Yousheng Cao	University of Pittsburgh
Roland W. Carlson	Lyndhurst. OH
Terry Carlton	Oberlin College
Kevin Carmichael	Wright State University
Walter H. Carnahan	Indiana State University
Delli Carpini	Boston University
Roger Carr	Stanford University
Leslie Carrice	CEI. Perry Power Plant
Christopher C. Carter	Ohio State University
Robert F. Cartland	Oberlin College
J. Castiglione	Case Western Reserve University
Mike Centanni	Ferro Corporation
Kevin Chaffee	Case Western Reserve University
Kenneth R. Challener	Cleveland. OH
B. S. Chandrasekhar	Case Western Reserve University
Jhy-Jiun Chang	Wayne State University
Mark Chantell	University of Dayton
Beverly G. Chantell	Dayton. OH
Stephen J. Cheesman	University of Toronto
Margaret Cheiky	Case Western Reserve University
Moukhtar Chemasani	University of Michigan
L. Chen	Case Western Reserve University
Hsiung Chen	Case Western Reserve University
Wei Chen	Wayne State University
Hong Chen	University of Toronto
Li Chen	Kent State University
Peijun Chen	University of Pittsburgh
Kai Chen	Wayne State University
Kan Chen	Ohio State University
Henry Hung-Chi Chen	University of Cincinnati
Yu Ming Chen	Michigan State University
Zi Ping Chen	Michigan State University
Jian H. Chen	Bowling Green State University
Zhidang Chen	York University
Ming Cheng	State University of New York. Buffalo
John C. Cheng	Monsanto Research Corporation
Gary Chottiner	Case Western Reserve University
Stan Christensen	Kent State University
C. W. Chu	University of Houston
Harry T. Chu	University of Akron
Andrew X. Chu	Lehigh University
Jean S. Chung	Ohio State University
Vince Cienciolo	University of Michigan
Robert Cieplechowicz	Wayne State University
Joseph Cihula	Case Western Reserve University
Ann Clapp	Willoughby. OH
Roger M. Clapp	Willoughby. OH
William Clark	Case Western Reserve University
Roy Clarke	University of Michigan

Brian Clasky	York University
David B. Cline	University of California, Los Angeles
T. Berry Cobb	Bowling Green State University
John R. Cogar	Kent State University
Roberto Colella	Purdue University
John W. Coltman	Pittsburgh, PA
Lin Cong	University of Cincinnati
Frances Cook	Fairview Park, OH
James Cook	Armco, Inc.
Al Cooper	Case Western Reserve University
Donald Cope	Sandusky High School
Judy Cortese	Shaker Heights, OH
Gregory Crawford	Kent State University
Robert Crawford	Avon High School
R. B. Creel	University of Akron
Keith R. Cromack	Ohio State University
Lijing Cui	State University of New York, Buffalo
Audrey Cullen	Wright State University
Frank H. Cverna	Los Alamos National Laboratory
Arnold Dahm	Case Western Reserve University
Liza D. Daley	Bennington College
Bill Dallas	Tucson, AZ
William B. Daniels	Newark, DE
Guy Danner	Case Western Reserve University
Debbie Dargo	Bowling Green State University
Suvro Datta	Carnegie-Mellon University
Pieter De Haseth	Case Western Reserve University
Partha Pratim Debroy	Carnegie-Mellon University
R. T. Deck	University of Toledo
Arthur Decker	NASA-Lewis Research Center
Richard DeFazio	Case Western Reserve University
Hendrik De Kruif	Meadville, PA
David DeLaney	Case Western Reserve University
Timothy Dennis	State University of New York, Buffalo
William A. Dent	Amherst, MA
Guy DeRose	Case Western Reserve University
K. DeVanand	Case Western Reserve University
Justiniano Diaz-Cruz	University of Michigan
Frank DiFilippo	Case Western Reserve University
Gioia Digiannantonio	Gates Mills, OH
Greg DiLisi	Case Western Reserve University
Peggy Holmes Dixon	Silver Spring, MD
J. William Doane	Kent State University
Mildred S. Dresselhaus	Massachusetts Institute of Technology
Robert Dunbar	Case Western Reserve University
G. Comer Duncan	Bowling Green State University
Kevin Dunphy	York University
D. Durand	Case Western Reserve University
Ryszard Duszak	York University
Abhijit K. Dutta	University of Pittsburgh
Gene Easter	Akron, OH
Thomas Eck	Case Western Reserve University
Bernard H. Eckstein	North Royalton, OH
David O. Edwards	Ohio State University
Alfred M. Eich	Cleveland, OH
Mostafa A. Elasaar	Kent State University
Richard Emmons	Kent State University
Kimberly Engle	Case Western Reserve University
Steve Eppell	Case Western Reserve University
John R. Erdmann	Kent State University
Howard Evans	University of Dayton

Howard Evans, Jr.	University of Dayton
Mark Everett	University of Toronto
Drew C. Fair	Kent State University
Yizhong Fan	University of Cincinatti
Yi Bing Fan	Michigan State University
Lillian Faro	University of Pittsburgh
David Farrell	Case Western Reserve University
Nicholas Fatica	Shaker Heights, OH
Sherwood L. Fawcett	Columbus, OH
An Feng	Notre Dame University
Henry Fenichel	University of Cincinnati
John Ferrante	NASA-Lewis Research Center
Bayard Keith Fetler	Carnegie-Mellon University
William Fickinger	Case Western Reserve University
Nelson Filho	Michigan State University
Mike Finley	Western Kentucky University
Michael E. Fisher	University of Maryland
Lorna Fleck	Youngstown State University
Leslie L. Foldy	Case Western Reserve University
Ian A. Folkins	University of Toronto
Allen H. Ford	Cleveland, OH
Jeffrey A. Fortner	Pennsylvania State University
John Foster	Case Western Reserve University
Wayne Fowler	University of Dayton
Julie E. Franklin	Ohio State University
Christopher Friedl	University of Cincinnati
Barbara Frisken	Toronto, ON
William R. Frisken	York University
Klaus Fritsch	John Carroll University
Edward Fritz	Cleveland, OH
Glenn Frye	Case Western Reserve University
Lewis P. Fulcher	Bowling Green State University
Mike J. Fuller	Geneva, OH
Gary G. Gadzia	Warren, OH
Dan Galehouse	Kent State University
Kenneth Gall	Boston University
Patrick Gallagher	University of Pittsburgh
Elizabeth Gallas	Michigan State University
Michael R. Gallis	Pennsylvania State University
Rohit Gangwar	Carnegie-Mellon University
Robert Garisto	University of Michigan
Carole Garrison	University of Akron
Fred E. Garry	General Electric Co.
Evalyn Gates	Case Western Reserve University
Jim Gauntner	NASA-Lewis Research Center
Daniel Gauntner	NASA-Lewis Research Center
Murray Gell-Mann	California Institute of Technology
Richard W. Genberg	Chagrin Falls, OH
Kenneth Gerhart	Avon High School
Mark Gertner	York University
Ivar Giaever	General Electric Company
Charles Giamati	NASA-Lewis Research Center
Dave Gillman	Falls Church, VA
David Glover	Case Western Reserve University
Daniel W. Goetz	Lima, OH
Paul A. Gohman	Wright State University
John P. Golben	Ohio State University
Maurice Goldhaber	Brookhaven National Laboratory
Minzhuan Gong	Michigan State University
Wen Guang Gong	Michigan State University
Jean C. Gordon	Lyndhurst, OH

William L. Gordon	Case Western Reserve University
Fred Gram	Cuyahoga Community College
Chris Graney	University of Dayton
John L. Grangaard	Newark, OH
Jeanette G. Grasselli	BP America
Tim Grayson	University of Dayton
David H. Green	Shaker Heights, OH
Paul Greenberg	NASA-Lewis Research Center
Lavina Greene	Lakewood, OH
Arthur F. Greene	Lakewood, OH
Ivan A. Greenwood	Stamford, CT
Doreen Grener	Avon High School
C. Frank Griffin	University of Akron
David F. Griffing	Miami University
Ross W. Groom	University of Toronto
Duane Gruber	University of California, San Diego
Gujrati	University of Akron
H. Richard Gustafson	University of Michigan
Jorge A. Gutierrez	Ohio University
Mark Haacke	Case Western Reserve University
Timothy Haas	Cleveland Heights, OH
Alec Habig	Wright State University
Jeanne F. Hagan	Cleveland, OH
Fred Hainsworth	Ryerson Polytechnical Institute
Paul D. Hambourger	Cleveland State University
Wallace Hamilton	Chagrin Falls, OH
William Hammond	Michigan State University
Jim Hammons	Fort Wayne, IN
Stephen Z. Hanzely	Youngstown State University
John George Hardie	University of Pittsburgh
Avaroth Harindranath	Ohio State University
Todd Harris	Case Western Reserve University
Michael J. Harrison	Michigan State University
Hans Haubold	Zentral Institut fur Astrophysik (Potsdam)
Mark Haugen	Purdue University
Edward J. Haugland	Cleveland Heights, OH
Ronald M. Haybron	Cleveland State University
George E. Haynam	Eastern Michigan University
Dov Hazony	Case Western Reserve University
Hui He	University of Michigan
Nancy Hecker	University of Michigan
Franz Heider	University of Toronto
Olle G. Heinonen	Case Western Reserve University
Carl Helrich	Goshen College
Lucy Henderson	University of Pittsburgh
A. James Henderson	Shaker Heights, OH
Robert C. Hendricks	NASA-Lewis Research Center
Thomas J. Hendrickson	Gettysburg College
P.N. Henriksen	University of Akron
Robert H. Henscheid	Clark Technical College
Daniel L. Herrick	Mentor, OH
Carl H. Hess	Mentor-on-the Lake, OH
A. H. Heuer	Case Western Reserve University
Edwin R. Hill	Chardon, OH
Robert Hinebaugh	Newark Air Force Base
George Hinshaw	Case Western Reserve University
Eric Hintz	Case Western Reserve University
James L. Hock	Sylvania Northview High School
Thomas Hockey	College of Wooster
Carol Hodanbosi	Barberton High School
Robert Hofstadter	Stanford University

Keith R. Honey	West Virginia Institute of Technology
Sibley W. Hoobler	Cleveland, OH
Alfred Hood	Warren, OH
Charles Horton	Case Western Reserve University
Billy M. Horton	Shaker Heights, OH
T. William Houk	Miami University
Pavel Hrma	Case Western Reserve University
Weimin Hu	Notre Dame University
Lujia Huang	Case Western Reserve University
Jing Huang	Kent State University
Qinqgi Huang	Bowling Green State University
Juyang Huang	State University of New York, Buffalo
Yi Yun Huang	Michigan State University
Stephen Hulbert	Case Western Reserve University
Dorothy Humel Hovorka	Shaker Heights, OH
Ling-Wai Hung	University of Toronto
Earle R. Hunt	Ohio University
James L. Hunt	University of Guelph
Daniel Huston	Marshall College
Lisa Hutzler	Bethany College
Darrell O. Huwe	Ohio University
Robert Hyland	Westlake, OH
H. Ikeda	University of Akron
Ruth Ilan	Case Western Reserve University
M. Imaeda	Solon, OH
Ralph Isaacs	Cincinnati, OH
Didarul Islam	Ohio University
Joseph Isler	University of Cincinnati
Conway G. Ivy	Moreland Hills, OH
Steven Izen	Case Western Reserve University
Don T. Jacobs	College of Wooster
Thomas Jacobson	NASA-Lewis Research Center
A. Edward Jaworowski	Wright State University
Doug Jayne	Case Western Reserve University
T. L. Jenkins	Case Western Reserve University
Wayne Jennings	Case Western Reserve University
Rob Jennings	Avon High School
Richard A. Jerdonek	Brecksville, OH
Yimin Ji	University of Akron
Hong Wen Jiang	Case Western Reserve University
Chen Jiang	University of Cincinnati
Yifeng Jiang	Ohio State University
Ping Jin	Eastern Michigan University
Yanhe Jin	Ohio University
Wei Jin	Michigan State University
Mark Alfred Joensen	Carnegie-Mellon University
Porter W. Johnson	Illinois Institute of Technology
Thomas L. Johnson	Ferris State College
William Johnson	Westminister College
De-Shien Jong	University of Cincinnati
Sheila Raja Jordan	University of Pittsburgh
Steven Joseph	Kent State University
Glenn M. Julian	Miami University
Menelaos S. Kafkalidis	Ohio State University
Michael Kahana	Case Western Reserve University
Jonghyun Kahng	Ohio State University
Hal Kalechafsky	University of Pittsburgh
Y. Kamutani	Lyndhurst, OH
Ho-Shik Kang	Ohio State University
Hua Kang	State University of New York, Buffalo
Thomas A. Kaplan	Michigan State University

Norman Kaplan	CALCOL Inc.
Betina Kapp	University of Pittsburgh
Wayne E. Karberg	Cleveland Electric Illuminating Co.
Mrunalini Karmarkar	Wayne State University
Miron Kaufman	Cleveland State University
Alex Kaufman	South Euclid, OH
Richard Kay	Malabar High School
Alexander Kaziner	Case Western Reserve University
Max J. Keck	John Carroll University
Joseph F. Keithley	Keithley Instruments, Inc.
Karsten Kell	Case Western Reserve University
Melinda A. Keller	Lakewood, OH
Scudder D. Kelvie	Hudson, OH
Alberta L. Kelvie	Hudson, OH
William R. Kendra	Wickliffe, OH
John Keyser	Silver Spring, MD
Mark Kief	Pennsyvania State University
Dug Young Kim	Notre Dame University
Yeoung Duk Kim	Michigan State University
Young-Hwan Kim	Case Western Reserve University
Paul A. Kimoto	Oberlin College
Michael J. King	BP America
David Kirkby	University of Toronto
Robert Kirshner	Harvard-Smithsonian Center for Astrophysics
Leonard S. Kisslinger	Carnegie-Mellon University
Richard Kistruck	Case Western Reserve University
Leroy Klein	Cleveland Heights, OH
Robert Klein	Cleveland State University
Jon J. Klement	BP America
Robert Klepfer	Case Western Reserve University
George Klinich	Case Western Reserve University
Gilles Klopman	Case Western Reserve University
Phil Klunzinger	University of Akron
Ron Knowlden	Western Kentucky University
Charles A. Knox	Case Western Reserve University
Jamie T. Knue	Cleveland, OH
Yoshinori Kobayashi	Case Western Reserve University
Edith Kobler	Oberageri, Switzerland
Richard Kobler	Oberageri, Switzerland
Ken Koehler	University of Cincinnati
Andrew Kolody	York University
Xian-Jun Kong	Kent State University
Daniel Koppel	Bowling Green State University
Greg Kostick	Case Western Reserve University
K. L. Kowalski	Case Western Reserve University
Jan Kozel	University of Toronto
Isay Krainsky	NASA-Lewis Research Center
Kevin Kratt	Avon High School
David Krus	Case Western Reserve University
Richard Krygowski	Youngstown, OH
Cindy Krysac	University of Toronto
Prasun K. Kundu	Ryerson Polytechnical Institute
Robert Kusner	Case Western Reserve University
T. J. Kvale	University of Toledo
Graciela Lacueva	John Carroll University
Peter Lagerlof	Case Western Reserve University
Al Lang	Bay Village, OH
Willis E. Lamb	University of Arizona
Walter Lambrecht	Case Western Reserve University
Francisco Lamelas	University of Michigan
Ken Lamkin	Western Kentucky University

Abbas Lamouri	Case Western Reserve University
Christian T. Lant	NASA-Lewis Research Center
William B. LaPlace	Cleveland, OH
Gregory Latta	Kent State University
Joseph Lawrence	Columbia City, IN
Abdelhamid Layadi	Carnegie-Mellon University
Sarah Leach	Elkhart, IN
Leon M. Lederman	Fermi National Accelerator Laboratory
Choon Lee	Case Western Reserve University
Jeongho Lee	Michigan State University
Hack Sung Lee	University of Pittsburgh
Ju-young Lee	Ohio State University
Joseph Y. Lee	Case Western Reserve University
Michael A. Lee	Kent State University
Harvey S. Leff	California State Polytechnic Institute
Tim Lenane	University of Michigan
Daniel J. Lesco	NASA-Lewis Research Center
Zili Li	ase Western Reserve University
Hong Li	Ohio State University
Xiaodong Li	Ohio University
Y. Li	Michigan State University
Hong Li	University of Pittsburgh
Gang Li	University of Michigan
Weixiong Li	Carnegie-Mellon University
Ming Li	State University of New York, Buffalo
Weijian Li	State University of New York, Buffalo
Ying Quan Liang	Michigan State University
Albert J. Libchaber	University of Chicago
G. Samuel Lightner	Westminister College
Chung-Yi Lin	Ohio State University
Hefen Lin	Kent State University
Carolyn Lindstrom	Shaker Heights, OH
Walter W. Lindstrom	Shaker Heights, OH
Milton Lipnick	Silver Spring, MD
Wenzhong Liu	University of Michigan
Zu-Wei Liu	Notre Dame University
Jiang Liu	Carnegie-Mellon University
Xichun Liu	State University of New York, Buffalo
D. M. Livingston	Wainscott, NY
Ikai Lo	State University of New York, Buffalo
James A. Lock	Cleveland State University
Kenneth F. Loje	Columbia, MD
Limin Lu	University of Pittsburgh
Yong Lu	Ohio State University
Carol Lucey	Jamestown Community College
Siu Yim Luie	University of Pittsburgh
Kwok-Hing Luk	Ohio State University
Jonathan Allen Lukin	Carnegie-Mellon University
Wayne Lundberg	Wright State University
Steven Lutgen	Case Western Reserve University
Bryan A. Luther	Ohio State University
Thomas E. Lynch	Gould Incorporated
Sergio Machado	College of Wooster
Stefan Machlup	Case Western Reserve University
William J. MacIntyre	Cleveland, OH
Innes K. MacKenzie	University of Guelph
Philip A. Macklin	Miami University
Michael MacMillan	University of Pittsburgh
Richard Madey	Kent State University
John R. Magan	West Virginia State College
S.D. Mahanti	Michigan State University

Nicholas Maina	College of Wooster
Marjorie Malley	State University of New York, Buffalo
Sidharth Mande	Ohio University
Mark Manley	Kent State University
J. Adin Mann, Jr.	Case Western Reserve University
Ralph Marinelli	Royal Oak, MI
Mark Marjewin	Seven Hills, OH
Leon F. Marker	Gencorp, Inc.
Michael Martens	Case Western Reserve University
Paul Martin	Mt. Sinai Hospital
Sean Mattingly	Andrews University
Margarita Mattingly	Andrews University
Robert McCarthy	Kent State University
Robert McCarthy, Jr.	Kent State University
Joseph McDermott	Ohio State University
Edward McDonnell	Brooklyn High School
Karen McEwen	College of Wooster
John D. McGervey	Case Western Reserve University
Michael J. McGettrick	Notre Dame University
Michael McGuigan	Rockefeller Institute
Bill McKeown	University of Cincinnati
Ariane McKiernan	University of Michigan
Robert S. McMichael	Ohio State University
Freddy Mdabane	College of Wooster
Gerald Mearini	Case Western Reserve University
Djamel Medjahed	University of Michigan
Horst Melcher	Zentral Institut fur Astrophysik, Potsdam
Bernardo Mendoza	State University of New York, Buffalo
H. W. Mergler	Case Western Reserve University
Sharon Metzger	Bowling Green State University
Edwin Meyer	Case Western Reserve University
R. G. Meyers	Euclid, OH
John Michel	Minneapolis, MN
Peter F. Michelson	Stanford University
Stephen Mihailov	York University
Jeffrey H. Miles	NASA-Lewis Research Center
Riley O. Miller	Lakewood, OH
Kris Miller	Western Kentucky University
Edward Millspaw	Geneva High School
D. Mimnagh	York University
M. S. M. Minhaj	Wayne State University
Felix Miranda	Case Western Reserve University
Vinod Mishra	Ohio State University
Dan Mocanu	Case Western Reserve University
Dieter Moeller	Case Western Reserve University
Thomas H. Monroe	NASA Lewis Research Center
Carlos V. Moreno	Ohio University
Michael Morgan	York University
Dan Morilak	Euclid, OH
David Moroi	Kent State University
Stephen W. Morris	University of Toronto
Anne Mowery	Bay High School
Mary Rose Mullin	Villa Maria, PA
Paul J. Murphy	Ohio State University
Tom Murphy	Western Kentucky University
R. Thomas Myers	Kent, OH
Phartaksarathy Nageswaran	Toledo, OH
Acotham D. Nair	Bowling Green State University
Hiroshi Nakanishi	University of Missouri
Kashi Nath	Case Western Reserve University
Tapan Nayak	Michigan State University

Guy Nechi	Cleveland, OH
John Nees	Rochester, NY
Marylyn Nelson	Enon Valley, PA
Robert C. Nerbun	University of South Carolina at Sumter
Nancy J. Nersessian	Princeton University
James G. Neville	Katy, TX
Andrew Newell	University of Toronto
Richard Newrock	University of Cincinnati
Frank A. Nezrick, Jr.	Fermi National Accelerator Laboratory
Leighann Nicholl	Michigan State University
William H. Nichols	John Carroll University
Kyoko Niedra	N. Ridgeville, OH
J. M. Niedra	NASA-Lewis Research Center
Carl E. Nielsen	Ohio State University
Anthony Noble	University of British Columbia
Garry E. Noe	Indiana University
Anne M. Nolan	General Electric Co., Richland, WA
William C. Nusbaum	Citation Computer Systems
Oddvar Nygaard	Case Western Reserve University
Ralph E. Oberly	Marshall University
Johnathan Obien	Wayne State University
Tom O'Donnell	University of Michigan
Warren B. Offutt	Eaton Corporation
Hisashi Ogata	University of Windsor
Koji Ohta	South Euclid, OH
Jeffery Olhoeft	Case Western Reserve University
Michael O'Mara	B. F. Goodrich Co.
Eric B. Omelian	Ohio University
Cheryl Ondrus	Avon High School
Ralph J. Oravec	Hampton Bays, NY
M. Ortalano	University of Pittsburgh
Mary Ann T. Orzech	Beavercreek, OH
Jay Oyster	Case Western Reserve University
Hyuk-Kyu Pak	University of Pittsburgh
William F. Palmer	Ohio State University
Frauke Palmer	Worthington, OH
Lakshmi Pandey	State University of New York, Buffalo
Sanjeev Pandey	University of Pittsburgh
Smio Pani	Marquette University
Wolfgang K. H. Panofsky	Stanford University
Thomas Paonessa	Case Western Reserve University
Marta Pardavi-Horvath	Ohio State University
Sonal Parikh	Wayne State University
Ta Ryeong Park	Michigan State University
David H. Parker	B. F. Goodrich Co.
Punit Parmananda	Ohio University
N. Parthasarthy	University of Toledo
R. K. Pathria	University of Waterloo
Edward L. Patrick	Dayton, OH
Myron R. Pauli	Navel Research Laboratories
Zoran Pazameta	State University of New York, Buffalo
N. Pearlman	Purdue University
Geoffrey Peddle	University of Toronto
Peter Pella	Gettysburg College
Larry Pelz	University of Dayton
Stephen Pepper	NASA-Lewis Research Center
Mark S. Perry	Wright State University
James R. Perry	Sheffield Lake, OH
Rajendra Persaud	York University
Bruce A. Peterson	Ohio State University
Frank Petiprin	Jamestown, NY

Valentin Petran	Case Western Reserve University
Rose Petrick	Case Western Reserve University
Hans O. Petsch	Storm Lake, IA
Rolfe Petschek	Case Western Reserve University
Charles Phillips	Silver Springs, MD
Philip Piltch	York University
Nancy Piltch	NASA-Lewis Research Center
Pirouz Pirouz	Case Western Reserve University
Alexander D. Pline	NASA-Lewis Research Center
Marie F. Plumb	Jamestown Community College
Paula A. Pomianowski	University of Pittsburgh
Ioana Popesch	University of Akron
James Porter	Eastern Michigan University
Richard Poruban	Avon High School
James E. Poth	Miami University
John Potthast	Van Buren, ME
Kenneth Powell	University of Michigan
Ashok Pradhan	Michigan State University
Bryan L. Preppernau	Ohio State University
Raymond J. Preski	Cuyahoga Falls, OH
Joseph R. Priest	Miami University
David G. Proctor	Baldwin-Wallace College
Bruce Pruitt	Western Kentucky University
Roger L. Ptak	Bowling Green State University
Agnar Pytte	Case Western Reserve University
Xiaodi Qi	University of Akron
Xiao Qiang	Notre Dame University
Rouzi Qiu	State University of New York, Buffalo
John C. Quail	Toledo, OH
Norman Ramsey	Harvard Unversity
Rex Ramsier	University of Akron
Phillip A. Ranney	Cleveland, OH
Narahari K. Rao	Ohio State University
P. S. Rao	Case Western Reserve University
Dipti P. Rath	Ohio State University
Anita R.B. Rauch	Ohio State University
Sarada Ravipati	Wright State University
Tane S. Ray	Jamaica Plain, MA
Steven C. Read	University of Toronto
Barbara J. Reeves	Ohio State University
Douglas Reilly	Los Alamos Scientific Laboratory
Edmunds Reineks	Case Western Reserve University
Frederick Reines	University of California, Irvine
Home Reitwiesner	Silver Spring, MD
John R. Reitz	Ann Arbor, MI
Christopher Rella	Case Western Reserve University
Jeff Remillard	University of Michigan
Shang-Yuan Ren	Notre Dame University
Steve Renfrow	Western Kentucky University
Jeff Renner	Western Kentucky University
Mario Renteria	Michigan State University
Eli Reshotko	Case Western Reserve University
Robert Rice	Case Western Reserve University
Bruce Richards	Case Western Reserve University
W. Bruce Richards	Oberlin College
Robert Riehemann	University of Cincinnati
Edward B. Riffle	BP America
Steven Risser	Kent State University
William Robertson	Case Western Reserve University
D. Keith Robinson	Case Western Reserve University

Roger W. Rollins	Ohio University
William B. Rolnick	Wayne State University
Darrell C. Romick	North Canton, OH
Katrina Rook	Carnegie-Mellon University
Charles Rosenblatt	Case Western Reserve University
Dale S. Ross	Michigan State University
Wojciech Rostafinski	NASA-Lewis Research Center
Richard Roth	Eastern Michigan University
Dan Rothstein	Kent State University
Bahram Roughani	University of Cincinnati
Jian-Zhong Ruan	Case Western Reserve University
Robert Rubinstein	Cleveland Heights, OH
John Runyan	Avon Lake High School
Mark Rupe	Enon Valley, PA
Marvin Russell	Western Kentucky University
Moti Rustgi	State University of New York, Buffalo
Om P. Rustgi	State University of New York, Buffalo
Douglas M. Rutan	Cleveland Heights, OH
Paul Rutt	Michigan State University
Edward H. Ryan	University of Akron
Robert Rynasiewicz	The Johns Hopkins University
Joseph Rynasiewicz	Lyndhurst, OH
Swapan K. Saha	University of Pittsburgh
Louis E. Sahr	General Motors Institute
Ali Salahieh	Case Western Reserve University
Imunahad Salahlekekonen	Beersheva, Israel
Claire Samson	University of Toronto
Jack Sanders	Broadview Heights, OH
Nicholas Sanduleak	Case Western Reserve University
Edward R. Sanford	Ohio University
Brad Sanford	Transylvania University
Denis Sankovic	Euclid, OH
Kumari Santosh	Case Western Reserve University
Richard Sasala	Bowling Green State University
Amitabh Satyam	Ohio State University
Majid Sawtarie	Bethany College
Joseph E. Sawyer	Kent State University
Samir I. Sayegh	Purdue University
Joachim Schambach	Kent State University
Michael J. Scharen	Kent State University
Arthur L. Schawlow	Stanford University
Xania Scheick	Case Western Reserve University
Hugo R. Schelin	Michigan State University
Ian Schlifer	York University
Herbert Schlosser	Cleveland State University
Hubert H. Schneider	Cleveland, OH
A. Benedict Schneider	Cleveland, OH
Paul D. Scholten	Miami University
Kristen Scholz	Avon High School
Robert Schroeder	Avon, OH
Donald Schuele	Case Western Reserve University
Thomas R. Schuerger	Monroeville, PA
John M. Schutter	Cleveland, OH
Pierre Schwaar	Monroe, CT
Leroy B. Schwartz	University Heights, OH
G. E. Schwarze	NASA-Lewis Research Center
Denise Schweinberg	Avon High School
Don Scipione	Cleveland Heights, OH
Richard G. Seasholtz	NASA-Lewis Research Center
Jonathan Secaur	Kent State University
Desmond Seekola	Kent State University

Benjamin Segall	Case Western Reserve University
Emilio Segre	University of California, Berkeley
E. J. Seldin	Baldwin-Wallace College
Wesley L. Shanholtzer	Marshall University
Natthi Sharma	Eastern Michigan University
Ananda Shastri	Carnegie-Mellon University
Thomas P. Sheahen	Western Technology, Inc.
Amy E. Shell	University of Michigan
Wei-Dian Shen	Wayne State University
Jun Shen	Notre Dame University
Paul Shepard	University of Pittsburgh
Ralph Sherriff	University of Pittsburgh
Russell Shew	Silver Spring, MD
Lie Shi	University of Michigan
Patrick J. Shields	Ohio State University
Yoseph Shiferaw	University of Cincinnati
Sugie Shim	Ohio State University
Joseph Daniel Shindler	Carnegie-Mellon University
Neil Shoemaker	Case Western Reserve University
William S. Shore	Cleveland, OH
Anthony Shoup	University of Cincinnati
Joseph Shovlin	Case Western Reserve University
Erwin F. Shrader	E. Cleveland, OH
Jiann-Shing Shyu	University of Cincinatti
Rezaul K. Siddique	University of Cincinnati
Robert Simha	Case Western Reserve University
Joseph Simon	Michigan State University
Peter M. Sinclair	University of Toronto
Edgar B. Singleton	Bowling Green State University
Stephen A. Siwecki	Wright State University
Richard Skellen	Case Western Reserve University
Bradley A. Smedley	Euclid, OH
Peter Smilovits	Beechwood, OH
Richard S. Smith	University of Toronto
Sandra Smith	Wayne State University
Douglas Smith	Cuyahoga Falls, OH
David A. Smith	Virginia Highlands Community College
John W. Snider	Miami University
Joseph L. Snider	Oberlin College
John E. Sohl	Ohio State University
Kelly Sommers	Avon High School
Upul Sonnadara	University of Pittsburgh
Elizabeth Sornsin	Case Western Reserve University
Jack A. Soules	Cleveland State University
Kenneth A. Soxman	Southwest Missouri State
Robert Speers	Firelands College
David Speiser	University of Louvain
P. Sreekanth	University of Toledo
R. Anthony Stallwood	Case Western Reserve University
Carol Stalzer	Firelands College
Mark Stan	Case Western Reserve University
Samuel Stansfield	University of Toledo
Gordon Starr	Case Western Reserve University
Robert Steagall	Case Western Reserve University
Thomas Steel	York University
Lindsay Steele	Kent State University
R. Steinberg	Berea, OH
Eric Steinfelds	Kent State University
Leon Sterling	Case Western Reserve University
Albert B. Stewart	Wright State University
Robert G. Stieglitz	Naval Research Laboratories

Thomas G. Stinchcomb	Chicago, IL
Mary Stockard	Avon High School
Mark Stollberg	Sterling Heights, MI
Robert Stone	University of Guelph
Ronald E. Stoner	Bowling Green State University
Ronald A. Strauss	Cleveland Heights, OH
Harry E. Stumpf	Carnegie-Mellon University
John Stupica	Broadview Heights, OH
Dan Styer	Oberlin College
Ravi Subramanian	Kent State University
Sridevi Subramanian	University of Toledo
Jason Sudy	Avon High School
Chijun Sun	Michigan State University
Narayan Sureswaran	Ohio State University
Bruce Sutherland	University of Toronto
Dedi Suyanto	University of Cincinnati
Kathryn Svinarich	Wayne State University
Karl D. Swartz	San Antonio, TX
Gene M. Szuch	Pepper Pike, OH
Masswood Tabib-Azar	Case Western Reserve University
Peggy Talley	Wayne State University
Chui Ling Tam	Michigan State University
Nasser Tamimi	Kent State University
Frank Tangherlini	College of the Holy Cross
Smio Tani	Marquette University
David Tarasick	York University
Michael A. Tartaglia	Michigan State University
Philip Taylor	Case Western Reserve University
Ronald C. Taylor	JBA Associates
Dan Teng	Notre Dame University
M. Tenhover	BP America
Sarath Tennakoon	Bowling Green State University
Indira Thekkekere	Bowling Green State University
G. Peter Thiel	Chesterton, OH
Chris Thomas	Meadville, PA
Johnathan Thompson	Western Kentucky University
Marshall Thomsen	Eastern Michigan University
Ben Paul Thrams	Cuyahoga Falls, OH
Hans Thunander	Chesterland, OH
James Shaw Tzuu Tien	Lyndhurst, OH
James V. Tietz	Dow Corning Corporation
Nico Tjandra	Carnegie-Mellon University
Ki-Wing To	University of Pittsburgh
William Tobocman	Case Western Reserve University
Terry P. Toepker	Xavier University
David Tomanek	Michigan State University
David Tomberg	West Geauga High School
Ed Tompkin	Rochester, NY
Wayne Tompkin	Rochester, NY
Stanley S. Toncich	Highland Heights, OH
Penger Tong	University of Pittsburgh
Dan G. Tonn	College of Wooster
R. Tosh	University of Pittsburgh
John H. Tripp	SKF, Netherlands
Joseph Trivisonno, Jr.	John Carroll University
Gary Troha	Bethany College
Angeliki D. Tserepi	Ohio State University
Thomas Tupper	Hermitage, PA
Anthony Turkevich	University of Chicago
Leonid Turkevich	Case Western Reserve University
Ryan D. Tweney	Bowling Green State University

Sergio E. Ulloa	Ohio University
Terrence H. Vacha	Cuyahoga Community College
Frank Valencic	Richmond Heights, OH
Ioana Valeriu	Case Western Reserve University
John Valley	Borromeo College of Ohio
John Van Aalst	University of Toronto
John B. Van Zytveld	Calvin College
Desiderio Vasquez	Notre Dame University
Scott Veirs	Colorado Springs, CO
Val R. Veirs	Colorado College
Shile Venkatonara	College of Wooster
Tom Vezdos	Avon High School
Pieter von Herrmann	Cleveland State University
George Vourvopoulos	Western Kentucky University
Will Wagner	Wright State University
Julie Wagoner	Bay Village, OH
Glen Wagoner	Bay Village, OH
David A. Wah	Oberlin College
Robert A. Walch	Ohio State University
Jearl D. Walker	Cleveland State University
Jim Walker	North Canton, OH
David L. Wallach	Bridgeville, PA
Yaxin Wang	Case Western Reserve University
Ming Jer Wang	Case Western Reserve University
Kuilong Wang	Case Western Reserve University
Peinan Wang	Wayne State University
Zhiqiang Wang	Ohio State University
Yun Wang	State University of New York, Buffalo
Bob Ward	Lakewood, OH
Raymond J. Wasniak	South Euclid, OH
Pam Wearsch	Avon High School
Ron Webber	University of Cincinnati
Dhammika Weerasundara	University of Pittsburgh
Charles D. Weller	Shaker Heights, OH
William E. Wells	Miami University
Gail S. Welsh	Oberlin College
Paul Wendel	Kent State University
Wolfgang Wenzel	Ohio State University
Myra West	Kent State University, Stark Campus
Bill Wheaton	Altadena, CA
Donald White	Brecksville, OH
Allen White	Lakewood High School
David Whitehouse	Boston University
William J. Whitesell	Antioch College
Steven Wickert	Case Western Reserve University
Mark T. Widmer	Cleveland, OH
Donald R. Wiff	Gencorp, Inc.
Philip E. Wigen	Ohio State University
Ronald Wilhelm	Bowling Green University
David Will	University of Cincinnati
Ernest Williams	Wright State University
Wendell Williams	Case Western Reserve University
W. D. Williams	NASA-Lewis Research Center
Meredith J. Williams	Dow Corning Corporation
W. Williamson	University of Toledo
Kenneth G. Wilson	Cornell University
C. W. Wilson	University of Akron
Jeff Wilson	NASA-Lewis Research Center
Alpha E. Wilson	West Virginia Institute of Technology
Winston Win	Wayne State University
John G. Winans	State University of New York, Buffalo

Rob Windisman	University of Toronto
Paul J. Wolfe	Wright State University
Lincoln Wolfenstein	Carnegie-Mellon University
Phil Womble	Western Kentucky University
Samuel S. Wong	University of Toronto
G. Theodore Wood	Cleveland State University
Ian Woodbury	University of Toronto
Judith Woodruff	General Electric Co.
David Woods	Case Western Reserve University
Robert M. Woods	United States Department of Energy
Robert T. Woodworth	Chicago, IL
Eric G. Woolgar	University of Toronto
Nancy F. Wright	Oak Ridge National Laboratory
Jin-Zhong Wu	University of Cincinnati
Xizeng Wu	University of Cincinnati
Odette V. Wurzburger	Cleveland Heights, OH
Robert Wysong	Wright State University
Dimitrios Xenikos	Ohio State University
Tingkang Xia	Ohio State University
Wei Zhu Xia	Michigan State University
Jing Xiao	University of Michigan
Ping Xie	University of Michigan
Kezhou Xie	State University of New York, Buffalo
Xu	Wayne State University
He Xu	Ohio State University
Guang Zhou Xu	Michigan State University
Hongming Xu	Michigan State University
Zhequn Xu	York University
Hong Yan	Michigan State University
Hai-tian Yang	Kent State University
Ernest Yeager	Case Western Reserve University
Noel K. Yeh	State University of New York at Binghamton
Arthur M. Yelon	Ecole Polytechnique, Montreal
Granddon Yen	Ohio State University
Qi Yin	University of Cincinnati
Qin-Yun Ying	State University of New York, Buffalo
Muturgesu Yoganathan	University of Pittsburgh
Elizabeth York	Lorain, OH
Minzi Yu	Case Western Reserve University
You-Yao Yu	State University of New York, Buffalo
Haiji Yuan	Kent State University
Ching Yue	Bowling Green State University
William Zacharias	Venture Lighting
Patrick Zajac	Kent State University
Eimad Zakariya	Case Western Reserve University
Edward J. Zampino	Garfield Heights, OH
Floyd Zehr	Westminister College
Xiao-cheng Zeng	Ohio State University
Rodney Zerkle	Port Clinton High School
Leping Zha	University of Pittsburgh
Renshi Zhang	Case Western Reserve University
Zheng Zhee Zhang	University of Pittsburgh
Jian Zhang	Ohio University
Jian-Kang Zhang	Ohio University
Lizeng Zhang	University of Cincinnati
Wei-ming Zhang	Kent State University
Tao Zhang	York University
Liang Zhao	Michigan State University
Jing Zhao	Michigan State University
E. Zhou	Kent State University
Bin Zhou	University of Pittsburgh

Fan Zhu	Michigan State University
Yongkang Zhu	Kent State University
Yang Zhu	Ohio State University
John Zilka	Shaker Heights, OH
Doris Zimmerman	Warren, OH
Joan Zinn	Cuyahoga Community College
Ivan Zitkovsky	University of Cincinnati
J. V. Zolotarevsky	Case Western Reserve University
Anne Zucker	North Royalton, OH
Fulin Zuo	Ohio State University
Tao Zuo	York University
Fredy Zypman	Case Western Reserve University

□□

AIP Conference Proceedings

		L.C. Number	ISBN
No. 1	Feedback and Dynamic Control of Plasmas – 1970	70-141596	0-88318-100-2
No. 2	Particles and Fields – 1971 (Rochester)	71-184662	0-88318-101-0
No. 3	Thermal Expansion – 1971 (Corning)	72-76970	0-88318-102-9
No. 4	Superconductivity in d- and f-Band Metals (Rochester, 1971)	74-18879	0-88318-103-7
No. 5	Magnetism and Magnetic Materials – 1971 (2 parts) (Chicago)	59-2468	0-88318-104-5
No. 6	Particle Physics (Irvine, 1971)	72-81239	0-88318-105-3
No. 7	Exploring the History of Nuclear Physics – 1972	72-81883	0-88318-106-1
No. 8	Experimental Meson Spectroscopy –1972	72-88226	0-88318-107-X
No. 9	Cyclotrons – 1972 (Vancouver)	72-92798	0-88318-108-8
No. 10	Magnetism and Magnetic Materials – 1972	72-623469	0-88318-109-6
No. 11	Transport Phenomena – 1973 (Brown University Conference)	73-80682	0-88318-110-X
No. 12	Experiments on High Energy Particle Collisions – 1973 (Vanderbilt Conference)	73-81705	0-88318-111–8
No. 13	π-π Scattering – 1973 (Tallahassee Conference)	73-81704	0-88318-112-6
No. 14	Particles and Fields – 1973 (APS/DPF Berkeley)	73-91923	0-88318-113-4
No. 15	High Energy Collisions – 1973 (Stony Brook)	73-92324	0-88318-114-2
No. 16	Causality and Physical Theories (Wayne State University, 1973)	73-93420	0-88318-115-0
No. 17	Thermal Expansion – 1973 (Lake of the Ozarks)	73-94415	0-88318-116-9
No. 18	Magnetism and Magnetic Materials – 1973 (2 parts) (Boston)	59-2468	0-88318-117-7
No. 19	Physics and the Energy Problem – 1974 (APS Chicago)	73-94416	0-88318-118-5
No. 20	Tetrahedrally Bonded Amorphous Semiconductors (Yorktown Heights, 1974)	74-80145	0-88318-119-3
No. 21	Experimental Meson Spectroscopy – 1974 (Boston)	74-82628	0-88318-120-7
No. 22	Neutrinos – 1974 (Philadelphia)	74-82413	0-88318-121-5
No. 23	Particles and Fields – 1974 (APS/DPF Williamsburg)	74-27575	0-88318-122-3
No. 24	Magnetism and Magnetic Materials – 1974 (20th Annual Conference, San Francisco)	75-2647	0-88318-123-1
No. 25	Efficient Use of Energy (The APS Studies on the Technical Aspects of the More Efficient Use of Energy)	75-18227	0-88318-124-X

No. 26	High-Energy Physics and Nuclear Structure – 1975 (Santa Fe and Los Alamos)	75-26411	0-88318-125-8
No. 27	Topics in Statistical Mechanics and Biophysics: A Memorial to Julius L. Jackson (Wayne State University, 1975)	75-36309	0-88318-126-6
No. 28	Physics and Our World: A Symposium in Honor of Victor F. Weisskopf (M.I.T., 1974)	76-7207	0-88318-127-4
No. 29	Magnetism and Magnetic Materials – 1975 (21st Annual Conference, Philadelphia)	76-10931	0-88318-128-2
No. 30	Particle Searches and Discoveries – 1976 (Vanderbilt Conference)	76-19949	0-88318-129-0
No. 31	Structure and Excitations of Amorphous Solids (Williamsburg, VA, 1976)	76-22279	0-88318-130-4
No. 32	Materials Technology – 1976 (APS New York Meeting)	76-27967	0-88318-131-2
No. 33	Meson-Nuclear Physics – 1976 (Carnegie-Mellon Conference)	76-26811	0-88318-132-0
No. 34	Magnetism and Magnetic Materials – 1976 (Joint MMM-Intermag Conference, Pittsburgh)	76-47106	0-88318-133-9
No. 35	High Energy Physics with Polarized Beams and Targets (Argonne, 1976)	76-50181	0-88318-134-7
No. 36	Momentum Wave Functions – 1976 (Indiana University)	77-82145	0-88318-135-5
No. 37	Weak Interaction Physics – 1977 (Indiana University)	77-83344	0-88318-136-3
No. 38	Workshop on New Directions in Mossbauer Spectroscopy (Argonne, 1977)	77-90635	0-88318-137-1
No. 39	Physics Careers, Employment and Education (Penn State, 1977)	77-94053	0-88318-138-X
No. 40	Electrical Transport and Optical Properties of Inhomogeneous Media (Ohio State University, 1977)	78-54319	0-88318-139-8
No. 41	Nucleon-Nucleon Interactions – 1977 (Vancouver)	78-54249	0-88318-140-1
No. 42	Higher Energy Polarized Proton Beams (Ann Arbor, 1977)	78-55682	0-88318-141-X
No. 43	Particles and Fields – 1977 (APS/DPF, Argonne)	78-55683	0-88318-142-8
No. 44	Future Trends in Superconductive Electronics (Charlottesville, 1978)	77-9240	0-88318-143-6
No. 45	New Results in High Energy Physics – 1978 (Vanderbilt Conference)	78-67196	0-88318-144-4
No. 46	Topics in Nonlinear Dynamics (La Jolla Institute)	78-57870	0-88318-145-2
No. 47	Clustering Aspects of Nuclear Structure and Nuclear Reactions (Winnepeg, 1978)	78-64942	0-88318-146-0
No. 48	Current Trends in the Theory of Fields (Tallahassee, 1978)	78-72948	0-88318-147-9

No. 49	Cosmic Rays and Particle Physics – 1978 (Bartol Conference)	79-50489	0-88318-148-7
No. 50	Laser-Solid Interactions and Laser Processing – 1978 (Boston)	79-51564	0-88318-149-5
No. 51	High Energy Physics with Polarized Beams and Polarized Targets (Argonne, 1978)	79-64565	0-88318-150-9
No. 52	Long-Distance Neutrino Detection – 1978 (C.L. Cowan Memorial Symposium)	79-52078	0-88318-151-7
No. 53	Modulated Structures – 1979 (Kailua Kona, Hawaii)	79-53846	0-88318-152-5
No. 54	Meson-Nuclear Physics – 1979 (Houston)	79-53978	0-88318-153-3
No. 55	Quantum Chromodynamics (La Jolla, 1978)	79-54969	0-88318-154-1
No. 56	Particle Acceleration Mechanisms in Astrophysics (La Jolla, 1979)	79-55844	0-88318-155-X
No. 57	Nonlinear Dynamics and the Beam-Beam Interaction (Brookhaven, 1979)	79-57341	0-88318-156-8
No. 58	Inhomogeneous Superconductors – 1979 (Berkeley Springs, W.V.)	79-57620	0-88318-157-6
No. 59	Particles and Fields – 1979 (APS/DPF Montreal)	80-66631	0-88318-158-4
No. 60	History of the ZGS (Argonne, 1979)	80-67694	0-88318-159-2
No. 61	Aspects of the Kinetics and Dynamics of Surface Reactions (La Jolla Institute, 1979)	80-68004	0-88318-160-6
No. 62	High Energy e^+e^- Interactions (Vanderbilt, 1980)	80-53377	0-88318-161-4
No. 63	Supernovae Spectra (La Jolla, 1980)	80-70019	0-88318-162-2
No. 64	Laboratory EXAFS Facilities – 1980 (Univ. of Washington)	80-70579	0-88318-163-0
No. 65	Optics in Four Dimensions – 1980 (ICO, Ensenada)	80-70771	0-88318-164-9
No. 66	Physics in the Automotive Industry – 1980 (APS/AAPT Topical Conference)	80-70987	0-88318-165-7
No. 67	Experimental Meson Spectroscopy – 1980 (Sixth International Conference, Brookhaven)	80-71123	0-88318-166-5
No. 68	High Energy Physics – 1980 (XX International Conference, Madison)	81-65032	0-88318-167-3
No. 69	Polarization Phenomena in Nuclear Physics – 1980 (Fifth International Symposium, Santa Fe)	81-65107	0-88318-168-1
No. 70	Chemistry and Physics of Coal Utilization – 1980 (APS, Morgantown)	81-65106	0-88318-169-X
No. 71	Group Theory and its Applications in Physics – 1980 (Latin American School of Physics, Mexico City)	81-66132	0-88318-170-3
No. 72	Weak Interactions as a Probe of Unification (Virginia Polytechnic Institute – 1980)	81-67184	0-88318-171-1
No. 73	Tetrahedrally Bonded Amorphous Semiconductors (Carefree, Arizona, 1981)	81-67419	0-88318-172-X

No. 74	Perturbative Quantum Chromodynamics (Tallahassee, 1981)	81-70372	0-88318-173-8
No. 75	Low Energy X-Ray Diagnostics – 1981 (Monterey)	81-69841	0-88318-174-6
No. 76	Nonlinear Properties of Internal Waves (La Jolla Institute, 1981)	81-71062	0-88318-175-4
No. 77	Gamma Ray Transients and Related Astrophysical Phenomena (La Jolla Institute, 1981)	81-71543	0-88318-176-2
No. 78	Shock Waves in Condensed Matter – 1981 (Menlo Park)	82-70014	0-88318-177-0
No. 79	Pion Production and Absorption in Nuclei – 1981 (Indiana University Cyclotron Facility)	82-70678	0-88318-178-9
No. 80	Polarized Proton Ion Sources (Ann Arbor, 1981)	82-71025	0-88318-179-7
No. 81	Particles and Fields –1981: Testing the Standard Model (APS/DPF, Santa Cruz)	82-71156	0-88318-180-0
No. 82	Interpretation of Climate and Photochemical Models, Ozone and Temperature Measurements (La Jolla Institute, 1981)	82-71345	0-88318-181-9
No. 83	The Galactic Center (Cal. Inst. of Tech., 1982)	82-71635	0-88318-182-7
No. 84	Physics in the Steel Industry (APS/AISI, Lehigh University, 1981)	82-72033	0-88318-183-5
No. 85	Proton-Antiproton Collider Physics –1981 (Madison, Wisconsin)	82-72141	0-88318-184-3
No. 86	Momentum Wave Functions – 1982 (Adelaide, Australia)	82-72375	0-88318-185-1
No. 87	Physics of High Energy Particle Accelerators (Fermilab Summer School, 1981)	82-72421	0-88318-186-X
No. 88	Mathematical Methods in Hydrodynamics and Integrability in Dynamical Systems (La Jolla Institute, 1981)	82-72462	0-88318-187-8
No. 89	Neutron Scattering – 1981 (Argonne National Laboratory)	82-73094	0-88318-188-6
No. 90	Laser Techniques for Extreme Ultraviolt Spectroscopy (Boulder, 1982)	82-73205	0-88318-189-4
No. 91	Laser Acceleration of Particles (Los Alamos, 1982)	82-73361	0-88318-190-8
No. 92	The State of Particle Accelerators and High Energy Physics (Fermilab, 1981)	82-73861	0-88318-191-6
No. 93	Novel Results in Particle Physics (Vanderbilt, 1982)	82-73954	0-88318-192-4
No. 94	X-Ray and Atomic Inner-Shell Physics – 1982 (International Conference, U. of Oregon)	82-74075	0-88318-193-2
No. 95	High Energy Spin Physics – 1982 (Brookhaven National Laboratory)	83-70154	0-88318-194-0
No. 96	Science Underground (Los Alamos, 1982)	83-70377	0-88318-195-9

No. 97	The Interaction Between Medium Energy Nucleons in Nuclei – 1982 (Indiana University)	83-70649	0-88318-196-7
No. 98	Particles and Fields – 1982 (APS/DPF University of Maryland)	83-70807	0-88318-197-5
No. 99	Neutrino Mass and Gauge Structure of Weak Interactions (Telemark, 1982)	83-71072	0-88318-198-3
No. 100	Excimer Lasers – 1983 (OSA, Lake Tahoe, Nevada)	83-71437	0-88318-199-1
No. 101	Positron-Electron Pairs in Astrophysics (Goddard Space Flight Center, 1983)	83-71926	0-88318-200-9
No. 102	Intense Medium Energy Sources of Strangeness (UC-Sant Cruz, 1983)	83-72261	0-88318-201-7
No. 103	Quantum Fluids and Solids – 1983 (Sanibel Island, Florida)	83-72440	0-88318-202-5
No. 104	Physics, Technology and the Nuclear Arms Race (APS Baltimore –1983)	83-72533	0-88318-203-3
No. 105	Physics of High Energy Particle Accelerators (SLAC Summer School, 1982)	83-72986	0-88318-304-8
No. 106	Predictability of Fluid Motions (La Jolla Institute, 1983)	83-73641	0-88318-305-6
No. 107	Physics and Chemistry of Porous Media (Schlumberger-Doll Research, 1983)	83-73640	0-88318-306-4
No. 108	The Time Projection Chamber (TRIUMF, Vancouver, 1983)	83-83445	0-88318-307-2
No. 109	Random Walks and Their Applications in the Physical and Biological Sciences (NBS/La Jolla Institute, 1982)	84-70208	0-88318-308-0
No. 110	Hadron Substructure in Nuclear Physics (Indiana University, 1983)	84-70165	0-88318-309-9
No. 111	Production and Neutralization of Negative Ions and Beams (3rd Int'l Symposium, Brookhaven, 1983)	84-70379	0-88318-310-2
No. 112	Particles and Fields – 1983 (APS/DPF, Blacksburg, VA)	84-70378	0-88318-311-0
No. 113	Experimental Meson Spectroscopy – 1983 (Seventh International Conference, Brookhaven)	84-70910	0-88318-312-9
No. 114	Low Energy Tests of Conservation Laws in Particle Physics (Blacksburg, VA, 1983)	84-71157	0-88318-313-7
No. 115	High Energy Transients in Astrophysics (Santa Cruz, CA, 1983)	84-71205	0-88318-314-5
No. 116	Problems in Unification and Supergravity (La Jolla Institute, 1983)	84-71246	0-88318-315-3
No. 117	Polarized Proton Ion Sources (TRIUMF, Vancouver, 1983)	84-71235	0-88318-316-1

No. 118	Free Electron Generation of Extreme Ultraviolet Coherent Radiation (Brookhaven/OSA, 1983)	84-71539	0-88318-317-X
No. 119	Laser Techniques in the Extreme Ultraviolet (OSA, Boulder, Colorado, 1984)	84-72128	0-88318-318-8
No. 120	Optical Effects in Amorphous Semiconductors (Snowbird, Utah, 1984)	84-72419	0-88318-319-6
No. 121	High Energy e^+e^- Interactions (Vanderbilt, 1984)	84-72632	0-88318-320-X
No. 122	The Physics of VLSI (Xerox, Palo Alto, 1984)	84-72729	0-88318-321-8
No. 123	Intersections Between Particle and Nuclear Physics (Steamboat Springs, 1984)	84-72790	0-88318-322-6
No. 124	Neutron-Nucleus Collisions – A Probe of Nuclear Structure (Burr Oak State Park - 1984)	84-73216	0-88318-323-4
No. 125	Capture Gamma-Ray Spectroscopy and Related Topics – 1984 (Internat. Symposium, Knoxville)	84-73303	0-88318-324-2
No. 126	Solar Neutrinos and Neutrino Astronomy (Homestake, 1984)	84-63143	0-88318-325-0
No. 127	Physics of High Energy Particle Accelerators (BNL/SUNY Summer School, 1983)	85-70057	0-88318-326-9
No. 128	Nuclear Physics with Stored, Cooled Beams (McCormick's Creek State Park, Indiana, 1984)	85-71167	0-88318-327-7
No. 129	Radiofrequency Plasma Heating (Sixth Topical Conference, Callaway Gardens, GA, 1985)	85-48027	0-88318-328-5
No. 130	Laser Acceleration of Particles (Malibu, California, 1985)	85-48028	0-88318-329-3
No. 131	Workshop on Polarized ^{3}He Beams and Targets (Princeton, New Jersey, 1984)	85-48026	0-88318-330-7
No. 132	Hadron Spectroscopy–1985 (International Conference, Univ. of Maryland)	85-72537	0-88318-331-5
No. 133	Hadronic Probes and Nuclear Interactions (Arizona State University, 1985)	85-72638	0-88318-332-3
No. 134	The State of High Energy Physics (BNL/SUNY Summer School, 1983)	85-73170	0-88318-333-1
No. 135	Energy Sources: Conservation and Renewables (APS, Washington, DC, 1985)	85-73019	0-88318-334-X
No. 136	Atomic Theory Workshop on Relativistic and QED Effects in Heavy Atoms	85-73790	0-88318-335-8
No. 137	Polymer-Flow Interaction (La Jolla Institute, 1985)	85-73915	0-88318-336-6
No. 138	Frontiers in Electronic Materials and Processing (Houston, TX, 1985)	86-70108	0-88318-337-4
No. 139	High-Current, High-Brightness, and High-Duty Factor Ion Injectors (La Jolla Institute, 1985)	86-70245	0-88318-338-2

No. 140 Boron-Rich Solids
         (Albuquerque, NM, 1985)   86-70246   0-88318-339-0

No. 141 Gamma-Ray Bursts
         (Stanford, CA, 1984)   86-70761   0-88318-340-4

No. 142 Nuclear Structure at High Spin, Excitation,
         and Momentum Transfer
         (Indiana University, 1985)   86-70837   0-88318-341-2

No. 143 Mexican School of Particles and Fields
         (Oaxtepec, México, 1984)   86-81187   0-88318-342-0

No. 144 Magnetospheric Phenomena in Astrophysics
         (Los Alamos, 1984)   86-71149   0-88318-343-9

No. 145 Polarized Beams at SSC & Polarized Antiprotons
         (Ann Arbor, MI & Bodega Bay, CA, 1985)   86-71343   0-88318-344-7

No. 146 Advances in Laser Science–I
         (Dallas, TX, 1985)   86-71536   0-88318-345-5

No. 147 Short Wavelength Coherent Radiation:
         Generation and Applications
         (Monterey, CA, 1986)   86-71674   0-88318-346-3

No. 148 Space Colonization: Technology and
         The Liberal Arts
         (Geneva, NY, 1985)   86-71675   0-88318-347-1

No. 149 Physics and Chemistry of Protective Coatings
         (Universal City, CA, 1985)   86-72019   0-88318-348-X

No. 150 Intersections Between Particle and Nuclear Physics
         (Lake Louise, Canada, 1986)   86-72018   0-88318-349-8

No. 151 Neural Networks for Computing
         (Snowbird, UT, 1986)   86-72481   0-88318-351-X

No. 152 Heavy Ion Inertial Fusion
         (Washington, DC, 1986)   86-73185   0-88318-352-8

No. 153 Physics of Particle Accelerators
         (SLAC Summer School, 1985)
         (Fermilab Summer School, 1984)   87-70103   0-88318-353-6

No. 154 Physics and Chemistry of
         Porous Media—II
         (Ridge Field, CT, 1986)   83-73640   0-88318-354-4

No. 155 The Galactic Center: Proceedings of
         the Symposium Honoring C. H. Townes
         (Berkeley, CA, 1986)   86-73186   0-88318-355-2

No. 156 Advanced Accelerator Concepts
         (Madison, WI, 1986)   87-70635   0-88318-358-0

No. 157 Stability of Amorphous Silicon Alloy
         Materials and Devices
         (Palo Alto, CA, 1987)   87-70990   0-88318-359-9

No. 158 Production and Neutralization of Negative
         Ions and Beams
         (Brookhaven, NY, 1986)   87-71695   0-88318-358-7

No. 159   Applications of Radio-Frequency Power to
Plasma: Seventh Topical Conference
(Kissimmee, FL, 1987)   87-71812   0-88318-359-5

No. 160   Advances in Laser Science–II
(Seattle, WA, 1986)   87-71962   0-88318-360-9

No. 161   Electron Scattering in Nuclear and Particle
Science: In Commemoration of the 35th Anniversary
of the Lyman-Hanson-Scott Experiment
(Urbana, IL, 1986)   87-72403   0-88318-361-7

No. 162   Few-Body Systems and Multiparticle Dynamics   87-72594   0-88318-362-5

No. 163   Pion–Nucleus Physics: Future Directions
and New Facilities at LAMPF
(Los Alamos, NM, 1987)   87-72961   0-88318-363-3

No. 164   Nuclei Far from Stability: Fifth
International Conference
(Rosseau Lake, ON, 1987)   87-73214   0-88318-364-1

No. 165   Thin Film Processing and Characterization of
High-Temperature Superconductors   87-73420   0-88318-365-X

No. 166   Photovoltaic Safety
(Denver, CO, 1988)   88-42854   0-88318-366-8

No. 167   Deposition and Growth:
Limits for Microelectronics
(Anaheim, CA, 1987)   88-71432   0-88318-367-6

No. 168   Atomic Processes in Plasmas
(Santa Fe, NM, 1987)   88-71273   0-88318-368-4